S. HRG. 112–209

AVIATION FUELS: NEEDS, CHALLENGES, AND ALTERNATIVES

HEARING

BEFORE THE

SUBCOMMITTEE ON AVIATION OPERATIONS, SAFETY, AND SECURITY

OF THE

COMMITTEE ON COMMERCE, SCIENCE, AND TRANSPORTATION
UNITED STATES SENATE

ONE HUNDRED TWELFTH CONGRESS

FIRST SESSION

JULY 28, 2011

Printed for the use of the Committee on Commerce, Science, and Transportation

U.S. GOVERNMENT PRINTING OFFICE

72–410 PDF WASHINGTON : 2012

For sale by the Superintendent of Documents, U.S. Government Printing Office
Internet: bookstore.gpo.gov Phone: toll free (866) 512–1800; DC area (202) 512–1800
Fax: (202) 512–2104 Mail: Stop IDCC, Washington, DC 20402–0001

SENATE COMMITTEE ON COMMERCE, SCIENCE, AND TRANSPORTATION

ONE HUNDRED TWELFTH CONGRESS

FIRST SESSION

JOHN D. ROCKEFELLER IV, West Virginia, *Chairman*

DANIEL K. INOUYE, Hawaii
JOHN F. KERRY, Massachusetts
BARBARA BOXER, California
BILL NELSON, Florida
MARIA CANTWELL, Washington
FRANK R. LAUTENBERG, New Jersey
MARK PRYOR, Arkansas
CLAIRE McCASKILL, Missouri
AMY KLOBUCHAR, Minnesota
TOM UDALL, New Mexico
MARK WARNER, Virginia
MARK BEGICH, Alaska

KAY BAILEY HUTCHISON, Texas, *Ranking*
OLYMPIA J. SNOWE, Maine
JIM DeMINT, South Carolina
JOHN THUNE, South Dakota
ROGER F. WICKER, Mississippi
JOHNNY ISAKSON, Georgia
ROY BLUNT, Missouri
JOHN BOOZMAN, Arkansas
PATRICK J. TOOMEY, Pennsylvania
MARCO RUBIO, Florida
KELLY AYOTTE, New Hampshire
DEAN HELLER, Nevada

ELLEN L. DONESKI, *Staff Director*
JAMES REID, *Deputy Staff Director*
BRUCE H. ANDREWS, *General Counsel*
TODD BERTOSON, *Republican Staff Director*
JARROD THOMPSON, *Republican Deputy Staff Director*
REBECCA SEIDEL, *Republican Chief Counsel and Chief Investigator*

SUBCOMMITTEE ON AVIATION OPERATIONS, SAFETY, AND SECURITY

MARIA CANTWELL, Washington, *Chairman*
DANIEL K. INOUYE, Hawaii
BARBARA BOXER, California
BILL NELSON, Florida
FRANK R. LAUTENBERG, New Jersey
AMY KLOBUCHAR, Minnesota
TOM UDALL, New Mexico
MARK WARNER, Virginia
MARK BEGICH, Alaska

JOHN THUNE, South Dakota *Ranking Member*
JIM DeMINT, South Carolina
ROGER F. WICKER, Mississippi
JOHNNY ISAKSON, Georgia
ROY BLUNT, Missouri
JOHN BOOZMAN, Arkansas
PATRICK J. TOOMEY, Pennsylvania
DEAN HELLER, Nevada

CONTENTS

Page

Hearing held on July 28, 2011 1
Statement of Senator Cantwell 1
Statement of Senator Thune 2
Statement of Senator Warner 4
Statement of Senator Lautenberg 4
Statement of Senator Klobuchar 43
  Prepared statement ....

WITNESSES

Dr. Lourdes Maurice, Executive Director, Office of Environment and Energy, Office of Policy, International Affairs, and Environment, Federal Aviation Administration 6
  Prepared statement 7
Terry Yonkers, Assistant Secretary for Installations, Environment and Logistics, United States Air Force 12
  Prepared statement 13
Billy M. Glover, Vice President, Environmental Strategy and Aviation Policy, Boeing Commercial Airplanes 16
  Prepared statement 18
Tom Todaro, CEO, AltAir Fuels, LLC 22
  Prepared statement 24
Sharon Pinkerton, Senior Vice President of Legislative and Regulatory Policy, Air Transport Association of America, Inc. (ATA) 26
  Prepared statement 28
Judith Canales, Administrator, Rural Business Service, United States Department of Agriculture 46
  Prepared statement 47
Richard L. Altman, Executive Director, Commercial Aviation Alternative Fuels Initiative (CAAFI) 49
  Prepared statement 51
John Plaza, President and CEO, Imperium Renewables, Inc. 55
  Prepared statement 58

APPENDIX

Jim Rekoske, Vice President/General Manager, Renewable Energy and Chemicals, Honeywell/UOP, prepared statement 67
Response to written questions submitted to Dr. Lourdes Maurice by:
  Hon. John D. Rockefeller IV 68
  Hon. Maria Cantwell 69
Response to written questions submitted to Billy M. Glover by:
  Hon. John D. Rockefeller IV 70
  Hon. Maria Cantwell 70
Response to written question submitted to Tom Todaro by:
  Hon. John D. Rockefeller IV 71
  Hon. Maria Cantwell 71
Response to written question submitted to Sharon Pinkerton by:
  Hon. John D. Rockefeller IV 71
  Hon. Maria Cantwell 72
Response to written question submitted by Hon. Maria Cantwell to:
  Judith Canales 72
  Richard Altman 73
Response to written question submitted to John Plaza by:
  Hon. John D. Rockefeller IV 75

Page

Report dated September 2010, entitled "Fueling the Future Force—Preparing the Department of Defense for a Post-Petroleum Era" by Christine Parthemore and John Nagl, Center for a New American Security 75

AVIATION FUELS: NEEDS, CHALLENGES, AND ALTERNATIVES

THURSDAY, JULY 28, 2011

U.S. SENATE,
SUBCOMMITTEE ON AVIATION OPERATIONS, SAFETY, AND SECURITY,
COMMITTEE ON COMMERCE, SCIENCE, AND TRANSPORTATION,
Washington, DC.

The Subcommittee met, pursuant to notice, at 10:02 a.m. in room SR–253, Russell Senate Office Building, Hon. Maria Cantwell, Chairman of the Subcommittee, presiding.

OPENING STATEMENT OF HON. MARIA CANTWELL, U.S. SENATOR FROM WASHINGTON

Senator CANTWELL. Good morning.

The Senate Committee on Commerce, Science, and Transportation, Aviation Operations, Safety and Security Subcommittee will come to order.

I want to thank all our witnesses for being here today. And we have two panels that we're going to hear from—obviously, those involved in the industry and production and those who are the end users in industry who are supporting how we get an infrastructure built to support alternative aviation fuels.

I want to thank Senator Thune for being here and being part of this hearing this morning.

I also especially want to thank Mr. Torado, Mr. Plaza, and Mr. Glover, all from Washington State. So, thank you all for traveling here to be here.

Let me take a moment to comment on the current FAA situation. Our air, our Nation's air traffic control system is being held hostage, and it's not fair to the thousands of workers wanting to get back to work and on the job, and it's not fair to Americans. There's about $200 million of weekly revenue that is not being collected. And what we need is to move forward on getting a continuing of the FAA. We've done this about 20 times now, so I'm hoping my colleagues in the House will come to terms with moving this legislation—particularly since the FAA overall bill is so close to being done as well. A lot of issues have been resolved. So, I'm hopeful that this week we will be able to resolve both of those issues.

This hearing is focused in the importance of investing in a stable supply chain for aviation production of biofuels. With the rising cost of jet fuel and the thriving American biofuel industry, we have an opportunity to help aviation by keeping costs down for the future.

More importantly, production of green jet fuel will mean real economic growth and opportunity, not just in the United States but around the country.

For over the past 50 years, Jet A fuel is the benchmark fuel used by commercial airlines. It's the gold standard that meets the strict requirements necessary for safe aircraft operation over a broad range of air temperatures and pressures.

But today's Jet petroleum A isn't perfect. It contains sulphur, results in pollutants, particulate matter that harms surfaces and the air quality at airports. The recovery processing and combustion of Jet A fuel contributes to 2 percent of the global greenhouse gasses. And come January, flying to and from the EU will be subject to the Europeans' cap-and-trade regime.

And Jet A is subject to the same elevated volatile prices that we have seen in the oil and gas markets. In 2003, the average Jet A was 85 cents per gallon, after peaking at over $3 in 2008, and then it settled at $2.24 last year. Fuel costs now represent a significant percentage of an airline's operating expense. The upward trajectory of fuel prices, when combined with its price volatilities, make it difficult for airlines to enter into long-term fuel contracts.

And all of these downsides are just going to get more challenging in the future. The demand for air travel is projected to grow. Airlines will require access to more fuel at ever-increasing prices. And the airline sector's greenhouse emissions will increase, and more communities around airports will be classified by EPA as non-attainment areas, making things more challenging.

So, that's why this hearing is so important, and the work that many of you have been doing is so important—the development and adoption of alternative fuels, particularly green fuels. And expanding the total fuel supply and making sure that we reduce our carbon footprint are all important issues.

By replacing foreign oil with domestic sources of fuel, we will also be creating U.S. jobs. The production of green jet fuel will real economic opportunity. I know in my state, by creating an important industry—obviously, aerospace is already an important industry—investing in biofuels could lead to new, nearly 200,000 jobs and $37 billion of economic impact over the next 12 years if we make the right investments today.

So, I thank you all for being here at this important hearing. I'm looking forward to what you have to say as we move forward.

We obviously need to make sure that the supply chain and the delivery of this fuel can be done. The Air Force plans a very cost-competitive process for domestic fuel and, via this alternative blend by 2016. We're going to hear about that today. The Navy has also started ambitious goals. So, I look forward to getting into more depth on exactly how we're going to move forward on this important issue.

And now I'd like to turn it over to the Ranking Member, Senator Thune, for his opening remarks.

STATEMENT OF HON. JOHN THUNE, U.S. SENATOR FROM SOUTH DAKOTA

Senator THUNE. Thank you, Madam Chairwoman.

I'd like to thank Chairwoman Cantwell for holding today's hearing with me on such an important topic, the aviation fuel needs of our country.

Today, the United States imports nearly 65 percent of its petroleum. By 2030, we're on a course for importing 70 percent. I think it goes without saying that our aviation industry is highly dependent upon foreign sources of fuel.

Because of our lack of domestic fuel production we're routinely subjected to major price fluctuations that cause taxpayers, consumers and industry billions of dollars each year.

The Department of Defense estimates that a $1 rise in the price per barrel of oil equates to $130 million per year in their fuel costs. This year alone, DOD may have to pay an additional $1.5 billion in fuel costs, and DOD's projected fuel use remains at their current levels. In fact, late last month the Air Force requested that $261 million be reprogrammed to cover the rising cost of fuel.

Delta Airlines similarly estimates that their fuel bills will rise by 35 percent this year, equating to an additional $3 billion in additional fuel costs.

Even more concerning is the fact that some of the largest foreign oil reserves are held by unstable and unfriendly regimes. For example, Venezuela was recently tabbed as having the largest oil reserve in the world. Meanwhile, Iran's proven oil reserves were recently upgraded by 10.3 percent, for a total of 151.2 billion barrels.

I think everyone here today would agree that it's in America's vital interest to secure domestic alternative fuel sources, whether they're synthetic or biofuels. Ultimately, this will improve our national security, promote jobs throughout the country, reduce costs to consumers and to the taxpayers, and will better help protect our environment by reducing greenhouse gas emissions.

As we listen to our panels today, my hope is that they can shed some light on current U.S. developments in this growing industry, the challenges faced by the industry, and how we can reach the goal of implementing 100 percent renewable aviation fuel.

That said, it's important to recognize that alternative aviation fuels are still too expensive at their current levels of approximately $35 per gallon. To make alternative fuels viable for commercial aviation and the Department of Defense, we need to promote a commercial infrastructure that dramatically increases production and distribution so that these synthetic and biofuels can compete with traditional jet fuels.

We can help build this infrastructure by decreasing the amount of time required for improving and certifying new fuel sources, as well as allowing for long-term Government contracts. Both of these changes are key to encouraging private sector investment in the development of alternative aviation fuels.

As such, my hope is that all parties here today can do everything possible to help speed up the certification and approval process. Ultimately, I believe this will help the aviation sector reach the FAA-established goal of using 1 billion gallons of alternative jet fuel per year by the year 2018.

Last, I want to applaud the Department of Defense and the Air Force for their relentless pursuit of alternative fuels. What an incredible achievement the Air Force has done and made in certifying nearly 99 percent of their aircraft fleet to fly on a 50-50 blend.

Again, thank you, Chairwoman Cantwell, for holding this hearing today so that we can protect the American consumer while advancing the economic vitality of this great Nation.

Senator CANTWELL. Thank you very much, Senator Thune.

And, Senator Warner, did you have any opening statement you'd like to make?

STATEMENT OF HON. MARK WARNER, U.S. SENATOR FROM VIRGINIA

Senator WARNER. I'll just be very brief, Madam Chair. Thank you for holding the hearing. I concur with your interests and the Ranking Member's interests in moving forward on this.

I just want to make sure as we think about alternative fuels and next generation avionics that we recognize that there is a critically important role that NASA-Langley can play in this process. And we've been looking forward to working with the Chair and others on seeing if we can take that expertise and try to work to create a public-private relationship where there would be a Center of Excellence building on some of the research and work on alternative fuels being developed, I know in the Northwest and elsewhere; but also make sure this NASA-Langley facility could become that Center of Excellence—not just around next generation fuels, but next generation aviation design that will make our aircraft more efficient and effective.

And I appreciate the Chair's willingness to have this hearing.

Senator CANTWELL. Thank you.

Well, we'll go ahead and get started with our panel, unless the Senator from New Jersey has an opening statement to make.

Senator LAUTENBERG. May I?

Senator CANTWELL. Not to put the Senator on the spot, but——

Senator LAUTENBERG. I stopped for some synthetic fuel here.

[Laughter.]

STATEMENT OF HON. FRANK R. LAUTENBERG, U.S. SENATOR FROM NEW JERSEY

Senator LAUTENBERG. Forgive the moment.

Madam Chairwoman, thank you very much for holding this important hearing. We can't go, as we have been, over these years.

Senator LAUTENBERG. Senator Lautenberg, could you—OK, good. Go right ahead.

Senator LAUTENBERG. Oh, yes. OK.

And having the toxic emissions problem, the costs problem, and the availability problem.

Aviation's importance to the country can't be overstated. Planes not only allow us to move people and goods from one coast to another in a matter of hours; they also keep our economy moving forward.

The commercial aviation system—11 million jobs, generates more than $1 trillion to our economic activity.

But, we know that the airlines are expensive to operate, and these costs do get passed on to customers, with more fees and higher prices. And some of this stems from a dramatic rise in the fuel costs, as I think is widely known.

Last year our country's airlines—I don't have to tell you, but it's a shocking amount of money: $36 billion on buying fuel. Compare this to a decade ago, when the airlines spent $15 billion for substantially less quantity. I'm sorry. $15 billion for slightly more quantity. In other words, this vital industry is spending twice as much to buy less fuel than it once did.

So, this doesn't include the environmental costs, damages from air travel; aircraft engines emitting soot, pollution, can be deadly, are fast-growing wide-world contributors to global warming pollution which is heating our planet to dangerous levels.

It is clear the airlines need to use more alternative fuels to reduce the industry's reliance on oil, minimize the harm aviation causes on our environment.

And we find that not all of the aviation fuels are created equal. Fuels that are made from coal products produce significantly more emissions than standard jet fuel. Using coal as an aviation fuel is taking us in the wrong direction—toward dirtier fuels at a time when our environment is already under assault.

And that's why I'm pleased that after years of investment it appears that clean biofuels are ready for widespread use. Last month I attended the Paris Air Show, where two separated planes arrived after making the first transatlantic flights using clean biofuels. One of them left from Morristown, New Jersey, and was powered by green jet fuel which New Jersey-based Honeywell produces.

While the use of these fuels is essential, alternative fuels alone won't do enough to strengthen our critically important aviation industry, cut costs, and protect our planet. We also need to continue building our country's next generation air traffic control system, and develop more fuel efficient aircraft and engines.

The NextGen System will use state-of-the-art GPS technology to help planes chart more direct routes, reduce delays, and limit idling, all of which will cut—help cut emissions. And that's why I'm so disappointed that our friends on the other side are playing politics with the FAA reauthorization, causing nearly 650 people to be furloughed at our Tech Center in New Jersey, the FAA Tech Center, where so much of the NextGen research and development is taking place. And making aircrafts more fuel-efficient, whether by redesigning planes or improving engines, will also improve performance and minimize environmental impacts.

The bottom line is this: We all want the aviation industry to be strong. It's essential for the well-being of America—both for the sake of our economy, the loyal employees who rely upon the airlines for their livelihoods. But we've also got to protect the health of our planet and the pocketbooks of the American people, who shouldn't be nickel-and-dimed with more fees and higher prices.

So, Madam Chairwoman, thank you. I look forward to hearing from our panel about how we can work together to deal with skyrocketing fuel costs, reduce greenhouse gas emissions, build the NextGen aircraft control system, and keep costs low for consumers.

And I thank all of you for being here with us today.

Senator CANTWELL. Thank you, Senator Lautenberg.

And now we'll move to our panel. I want to welcome them.

Thank you very much for being here, and your testimony today, Dr. Lourdes Maurice, from the Environmental and Energy section of FAA; TerryYonkers, Assistant Secretary for USAF Installations; and Bill Glover, from, the Vice President of Environmental and Aviation Policy of Boeing; Tom Todaro, Chief Executive Officer of Targeted Growth and Alt. Air Fuels; and Sharon Pinkerton, who is the Senior Vice President of the Air Transportation Association.

Welcome to all of you. We look forward to hearing your testimony.

And we're going to start with you, Dr. Maurice.

STATEMENT OF DR. LOURDES MAURICE, EXECUTIVE DIRECTOR, OFFICE OF ENVIRONMENT AND ENERGY, OFFICE OF POLICY, INTERNATIONAL AFFAIRS, AND ENVIRONMENT, FEDERAL AVIATION ADMINISTRATION

Dr. MAURICE. Thank you, and good morning.

Madam Chairwoman, Senator Thune, and members of the Subcommittee, thank you very much for inviting me to testify before you.

Today, commercial aviation faces a number of challenges—fuel costs, environmental impacts, and energy security—that sustainable jet fuels can help address.

Fuels derived from biomass may offset a portion of the carbon produced by the aircraft, as well as mitigate air quality impacts from emissions of sulphur and particulate matter. And domestic alternatives to petroleum type fuel can expand and diversify jet fuel supplies and contribute to price stability and supply security.

Today's hearing is well-timed. Aviation continues to make enormous progress identifying, testing, and approving alternative jet fuels for commercial use.

As you may know, the FAA does not directly approve jet fuel. Rather, we approve aircraft to operate on fuel whose quality and safety is managed by industry-developed specifications. In partnership with industry, we have identified a number of drop-in fuels. These are alternative jet fuels that can replace petroleum jet fuel without the need to modify aircraft engines and fueling infrastructure.

On July 1, ASTM International, the industry standards organization, reached a major milestone with the announcement of the approval for use of a new class of jet biofuels at a 50 percent blend level with petroleum jet fuel. Known as HEFA, or hydroprocessed esters and fatty acids jet fuels, the biofuel component can be made from renewable plant oils. This approval was the product of over 3 years of collaboration by FAA, DOD, manufacturers, airlines and fuel suppliers.

Alternative jet fuels are a key component of the FAA's environmental energy approaches for NextGen. Over the past 5 years, the FAA has taken a comprehensive approach in cooperation with other stakeholders and enabled the end use of sustainable jet fuels in commercial jet aircraft. We have worked with our partners through CAAFI, who you will hear from, to address many of the issues related to creating drop-in sustainable jet fuels.

The FAA's role is multi-fold. It includes support of fuel properties and performance testing, facilitation of fuel approval by ASTM

International, conducting environmental measurements and analysis, and facilitating information exchange between stakeholders.

FAA has worked in partnership with other departments and agencies. For example, the collaboration with USDA has created a Feedstock Readiness Level Tool to help us determine the ability to use various agricultural or forest-based feedstocks to produce jet fuels.

The FAA's CLEEN program and other NextGen investments in environment and energy research, are vehicles we at the FAA are using to address the certification and environmental issues of alternative aviation fuels. We appreciate the Subcommittee's support for these efforts.

As we move forward, FAA recommends focusing on certain areas: We must foster the development and production scale-up of appropriate feedstocks for aviation biofuels. We must continue to support development, testing, and approval through ASTM International of additional classes of drop-in biofuels. We must quantify environmental impacts and understand how sustainability issues will be managed.

A major hurdle is the lack of jet biofuel infrastructure. The economic slowdown diminished the availability of capital to respond to the opportunities that aviation uniquely provides. However, we believe that successful production facilities can be built at locations which combine feedstock availability, and access to airports and U.S. airlines eager to use these new fuels.

The Nation has often counted upon the skills of the aerospace community to lead the way in technical innovation. Sustainable jet fuels offer the opportunity to team aerospace with agriculture, energy, and environmental communities to address the challenges we face.

Madam Chair and members of the Subcommittee, thank you so much. I started my career working in alternative fuels 28 years ago, and I am so impressed that the Senate is paying attention to this issue. So, thank you so much, and I welcome any questions that you may have.

[The prepared statement of Dr. Maurice follows:]

PREPARED STATEMENT OF DR. LOURDES MAURICE, EXECUTIVE DIRECTOR, OFFICE OF ENVIRONMENT AND ENERGY, OFFICE OF POLICY, INTERNATIONAL AFFAIRS, AND ENVIRONMENT, FEDERAL AVIATION ADMINISTRATION

Madam Chair, Senator Thune, and members of the Subcommittee:

Thank you for inviting me to testify before you today on "Aviation Fuels: Needs, Challenges, and Alternatives." I am the Executive Director of the Office of Environment and Energy for the Federal Aviation Administration (FAA). In that role, I also serve as the environmental team co-leader for the Commercial Aviation Alternative Fuels Initiative (CAAFI). I am pleased to speak to the Subcommittee today about the development and deployment of sustainable alternative jet fuels.

Today, commercial aviation faces a number challenges—fuel cost, environmental impacts and energy security—that sustainable jet fuels can help to address. Fuels that are derived from biomass may offset a portion of the carbon produced by the aircraft as well as mitigate air quality issues such as emissions of sulphur and particulate matter. And domestic alternatives to petroleum jet fuel can expand and diversify the jet fuel supply and contribute to price stability and supply security.

Industry, government and academia all need aviation to get these fuels off the drawing board and into the gas tank. Indeed, the Future of Aviation Advisory Committee, which was founded by Transportation Secretary LaHood in 2010, singled out aviation fuels and the environment in one of its recommendations.

I believe that today's hearing is well timed. Aviation continues to make enormous progress in identifying, testing, and approving alternative jet fuels for use by commercial airlines. As you may know, the FAA has the responsibility to make sure that any aircraft, aircraft engine or part, or fuel that is used in aviation is safe and performs to set standards. In partnership with industry, we have identified a number of alternative jet fuels (including sustainable jet fuels) that can replace petroleum jet fuel without the need to modify aircraft, engines, and fueling infrastructure. These are often referred to as "drop in" fuels. Drop-in fuels are a near-term solution to addressing aviation environmental and energy challenges, and enable us to maintain the existing commercial airline fleet.

The aviation sector is well positioned to adopt alternative fuels and is in fact beginning to do so.[1] Moreover, this effort is critical to achieving the level of environmental and energy performance that will allow sustained growth of the Nation's aviation system. FAA has set an aspirational target for use of 1 billion gallons of alternative jet fuel per annum by 2018.

Overview of FAA Role and Activities

Alternative jet fuels are a key component of the FAA's environmental and energy approaches for Next Generation Air Transportation System (NextGen). Over the past 5 years the FAA has taken a comprehensive approach, in cooperation with other departments and agencies, industry, and academia to address barriers, and enable the adoption, production, and end use of sustainable jet fuels in commercial jet aircraft. Beginning in 2006, we have worked with industry and government partners through CAAFI to address the business, research and development, environmental, and certification issues related to creating "drop-in" sustainable jet fuels for today's commercial aircraft.

The FAA's role has been multifold. It includes support of fuel properties and performance testing and demonstration; facilitation of fuel approval by the industry standard setting organization, ASTM International; conducting environmental measurements and analysis; and facilitating information exchange among industry and government stakeholders as a co-sponsor of CAAFI. FAA has worked in partnerships with the Department of Defense (DOD), the National Aeronautics and Space Administration (NASA), the Department of Energy (DOE), the Environmental Protection Agency (EPA), the Department of State (DOS), Department of Commerce (DOC), and the Department of Agriculture (USDA) to advance technical research and development, as well as environmental, fuel standard setting, and deployment efforts needed to support sustainable alternative fuels for jet aircraft.

The FAA's Continuous Lower Energy, Emissions and Noise (CLEEN) program, as well as NextGen investments in environment and energy research, are vehicles available to address the certification and environmental issues of alternative fuels. We appreciate the Subcommittee's support for these efforts.

Fuel Approvals

FAA does not directly approve jet fuel. Rather the FAA approves aircraft to operate on fuel whose quality and safety is managed by industry-developed specifications, such as ASTM International. FAA personnel and funding have, however, been crucial to facilitation of this specification development process at ASTM International. The ASTM alternative jet fuels standard (also known as Specification D7566) was first issued in September 2009 and at that time approved use of blends of up to 50 percent synthetic fuels made via the Fischer-Tropsch process, which produces synthetic fuels from feedstocks including coal, natural gas or biomass.[2] The specification is structured to allow for the addition of new fuels as they are qualified for use. The writing of the specification and its revisions are accomplished via a collaborative and consensus driven process that is facilitated by FAA's leadership of the CAAFI certification and qualification team.

On July 1, 2011, the aviation community reached a major milestone when ASTM International approved a revision of the D7566 specification to add alternative jet fuels made from bio-derived oils. Known as HEFA (hydroprocessed esters and fatty acids) jet fuels, they can be made from renewable plant oils such as camelina, jatropha, and algae or waste fats which are then mixed with petroleum jet fuel up to a 50 percent blend level. This represents the culmination of more than 3 years

[1] Following ASTM approval Lufthansa, KLM and UK airline Thompson Airways have begun regular commercial flights using HEFA biofuels sourced from Finnish fuel supplier Neste Oils (Lufthansa) and U.S. fuel supplier Dynamic Fuels (KLM, Thompson).

[2] The Fischer-Tropsch (FT) process created in Germany in the 1930s and later commercialized in South Africa by SASOL, produces synthetic fuels from any source of carbon and hydrogen via gasification and then conversion to fuels using chemical catalysts. Feedstocks include coal, natural gas or biomass (*e.g.*, crop residue, wood chips, or waste).

of collaborative work by FAA, DOD, and industry, including the engine and aircraft manufacturers, airlines, and fuel suppliers. The approval assures the safety and performance of the fuel and is enabling, for the first time, the commercial use of biofuel by airlines globally.

HEFA was the second alternative jet fuel to be approved for use by ASTM since 2009, but it will not be the last. Cooperative testing of additional advanced alternative jet fuels is already underway by FAA, DOD, and industry. From FAA's perspective, this is part of a strategic approach to approving as many commercially viable and environmentally sustainable alternative jet fuel options as possible.

Some of the fuel testing to support approval is being done through the FAA's CLEEN program. CLEEN supports maturation of green engine and airframe technologies and development and testing of alternative fuels. Under the CLEEN program, FAA leverages the Federal investment by partnering with industry.[3] For example, CLEEN has supported the Boeing Company to conduct aircraft fuel system materials compatibility testing of HEFA fuels. With Honeywell, we are testing the use of fully renewable jet biofuels. With Rolls Royce, we are doing fuel property, performance and engine testing to support evaluation of early stage, promising novel sustainable jet fuels.

Through the Department of Transportation/Research and Innovative Technology Administration's (DOT/RITA) Volpe National Transportation Systems Center (Volpe Center), the FAA will shortly be announcing grant awards to benchmark fuel quality control procedures, to conduct engine durability tests with alternative fuels, and to perform key testing to support qualification and certification of novel jet biofuels from alcohols, pyrolysis, and other processes. These are intended to support the next round of fuel approvals that are currently targeted to begin in 2013.

Environmental Assessment

In addition to certification and qualification of fuels, FAA is working to improve our understanding of the environmental benefits and impacts of alternative jet fuels. The U.S. has National Ambient Air Quality Standards for particulate matter emissions, and 44 percent of our 50 largest airports reside in areas of non-attainment. Common to all alternative fuels under consideration is their potential to reduce particulate matter emissions. Working with NASA, we have obtained direct measurements of in-service aircraft engines that clearly validate these benefits.

Through the Partnership for AiR Transportation Noise and Emission Reduction (PARTNER) Center of Excellence, FAA is funding assessments of emissions for alternative fuels including sustainable jet fuels.[4] The National Academies of Science's Airports Cooperative Research Program (ACRP) is supporting a project to understand the costs and the potential air quality benefits of alternative jet fuel use at commercial airports.

Reducing aviation's contribution to carbon dioxide emissions and climate change impacts are key potential benefits of alternative jet fuels. Measuring those benefits requires quantifying the full life cycle emissions from alternative fuel production, distribution, and operation. The FAA and the U.S. Air Force are jointly funding the development of greenhouse gas life cycle analyses (LCA) through the FAA's PARTNER Center of Excellence.[5] Results show that certain alternative jet fuels could realize CO_2 lifecycle reductions as high as 80 percent. We continue to work and consult with EPA, DOE and a team of researchers to improve and broaden these analyses. The CAAFI Environment team, which FAA co-leads, is similarly involved in coordinating a broad group of experts to look at sustainability questions such as water use, food versus fuel, and invasiveness to provide insight into how sustainability certification may be conducted. And, through Volpe Center grant awards mentioned above, the FAA will support evaluation of biofuel sustainability criteria.

Key Recent Developments

A review of recent developments will give you a sense of the tremendous momentum behind alternative jet fuels and demonstrate the broad industry and interagency cooperation and innovative partnerships that are providing the push.

---

[3] All CLEEN projects include a one to one cost share commitment by industry although the industry contribution leveraged is sometimes greater.

[4] This PARTNER project is *Emissions Characteristics of Alternative Aviation Fuels* and *Ultra Low Sulfur (ULS) Jet Fuel Environmental Cost Benefit Analysis.* More information about PARTNER is available at *http://web.mit.edu/aeroastro/partner/projects/index.html.*

[5] For work to develop alternative jet fuel life cycle analyses, see PARTNER Center of Excellence *Project 17: Alternative Jet Fuels* and *Project 28: Alternative Jet Fuel Environmental Cost Benefit Analysis* at *http://web.mit.edu/aeroastro/partner/projects/index.html.*

Jet Biofuels Approval and Flights

The July 1, 2011, ASTM International approval of HEFA alternative jet fuels made from bio-derived oils was a landmark. This has been followed by the first commercial service flights with HEFA biofuels by four airlines in Europe and has energized plans for possible production and fuel purchase agreements here in the United States.

Paris Air Show Alternative Aviation Fuels Showcase

In June 2011, the FAA and CAAFI worked with the Department of Commerce to showcase alternative jet fuel suppliers and U.S. and international airlines as a central event at the Paris Airshow. The event included visits of support by Secretary of Transportation Ray LaHood, FAA Administrator Babbitt, Acting Secretary of Commerce Sanchez, and Secretary of Agriculture Vilsack. It was successful in focusing the attention of the biofuels and agriculture communities and the media on the need and opportunity presented by aviation. Significant industry highlights at the airshow included the announcement by 7 U.S. airlines of negotiation with biofuel supplier Solena for 16 million annual gallons of fuel from waste in Northern California and two successful transatlantic biofuel flights to the airshow by Honeywell and Boeing.

U.S.—Brazil Partnership for the Development of Aviation Biofuels

During President Obama's visit to Brazil in March 2011, the United States and Brazil announced the creation of a "Partnership for the Development of Aviation Biofuels" under the Memorandum of Understanding between the United States and Brazil to Advance Cooperation on Biofuels signed on March 9, 2007. The FAA is a key participant and is engaged with the DOD, DOE, USDA, and other Federal departments and agencies to identify and carry out cooperative activities with Brazilian counterparts under this MOU. This agreement represents cooperation by the world's two largest biofuels producers and two important aviation States to support the development of sustainable jet fuels. It builds upon and will leverage existing collaboration with Brazil already underway via CAAFI.

FAA and USDA Partnership to Develop Renewable Jet Fuels

In October 2010, the FAA and the U.S. Department of Agriculture (USDA) signed a 5 year agreement that creates a framework of cooperation between FAA's Office of Environment and Energy, the USDA's Agricultural Research Service (ARS), and the USDA Office of Energy Policy and New Uses (OEPNU). Under the partnership, the three offices bring together their experience in research, policy analysis and air transportation to assess the availability of different kinds of feedstocks that will be needed by biorefineries to produce sustainable jet fuels. The collaboration has created the feedstock readiness level (FSRL)[6] tool, developed by the USDA and FAA to enable the determination of the stage of readiness of agricultural or forest-based feedstock for the production of commercial and military aviation biofuels. A public version is expected to be released soon.

Farm to Fly Partnership Formed between Airlines, USDA, and Boeing

In July 2010, the USDA joined with CAAFI sponsor Air Transport Association of America (ATA) and the Boeing Company in a resolution to "accelerate the availability of sustainable aviation biofuels in the United States, increase domestic energy security, and establish regional supply chains and support rural development." The agreement included the formation of a "Farm to Fly" working group that is identifying opportunities for accelerating a domestic jet biofuel production industry and supporting economic development in rural communities. This is a promising innovative effort that can further the interests of U.S. agriculture and U.S. aviation.

Challenges Ahead

To achieve the successful development and deployment of sustainable jet fuels in commercial aviation, we view the following areas as hurdles, as well as opportunities for future focus:

> We must foster the development and production of appropriate feedstocks for aviation biofuels. Expanding the number and availability of crops appropriate for jet fuel conversion and optimizing their production are necessary to reduce

[6] The *Feedstock Readiness Level* (FSRL) tool was developed by the USDA and FAA to enable the determination of the stage of readiness of agricultural or forest-based feedstock for the production of commercial and military aviation biofuels. The FSRL tool was structured to complement the *Fuel Readiness Level* (FRL) tool in use by the aviation industry. FSRL can be used to facilitate a coordinated allocation of resources to effectively develop a viable aviation biofuels industry.

costs, enable commercial deployment, and maintain sustainability. Our work with the USDA on the feedstock readiness level is a promising start, and we expect to continue to build on this collaboration.

We must continue to support the development, testing and approval of advanced biofuel conversion processes for high energy "drop in" hydrocarbon biofuels. Our past successes with Fischer-Tropsch and HEFA fuels would not have been possible without the leadership and contributions of the FAA, and this level of support must be maintained to move forward with new renewable and sustainable jet fuels. In addition to the CLEEN program and Volpe Center grant awards, the FAA resources will need to be allocated to support the ASTM International process to qualify and approve these new fuels. Investments by DOE, USDA, and DOD's Defense Advanced Research Projects Agency (DARPA) in these areas have been and will continue to be crucial. FAA must continue to work with DOD to coordinate the qualification and certification testing of both commercial and military fuels to make the best use of our limited resources.

The next hurdle is accurately quantifying environmental impacts. Assessments of both air quality and greenhouse gas life cycle emissions impacts must continue to be timely and thorough as new fuel options emerge. For example, FAA, in collaboration with EPA and NASA, needs to populate emissions prediction models with measured emissions data for emerging sustainable jet fuels. Acquiring such data is empirical in nature and requires significant testing and investment. Reducing the uncertainties associated with land use changes, fertilizer use, and impacts on the quality and quantity of water resources, greenhouse gas inherent in-life cycle analyses (that is, from harvest to processing to transport and use of the sustainable jet fuels) will also require significant effort and investment. The collaboration of all stakeholders involved is needed to ensure an agreeable and accurate framework. We must continue to facilitate defined national and international sustainability criteria and Life Cycle Analysis (LCA) methodologies to provide certainty and compatibility regarding how fuels will be judged and accepted.

The final hurdle is the lack of jet biofuel infrastructure investment by private industry. The economic slowdown diminished the ability and interest of conventional investment sources to respond to the opportunities that aviation uniquely provides. However, we believe that successful production facilities can be built with relatively modest investment at locations which combine feedstock availability, existing biofuel infrastructure, need for air quality gains, access to airports and U.S. airlines eager to use sustainable jet fuels. Progress being made by the Farm to Fly effort and via USDA, DOE and DOD programs suggest that early deployment may be close at hand, but will continue to require near term support.

Aviation's dependence on high-density liquid hydrocarbon fuels for the foreseeable future is perhaps unique. Unlike surface transportation, we won't have an electric option in the near future. Another unique characteristic of U.S. commercial aviation is concentrated fueling infrastructure, where 80 percent of all jet fuel is used in only about 35 locations, *i.e.,* at our busiest airports. Airports also provide an opportunity for distributing the co-products of sustainable jet fuel production (such as diesel) due to the many different fuel users on airports. The National Academies of Science's ACRP is sponsoring projects to assess the opportunity presented to airports of alternative fuel production and distribution. These realities of dependence and concentrated infrastructure should lead to aviation becoming a "first mover" in the deployment of alternative fuels. A final plus is the enthusiasm and commitment of the aviation industry to pioneer sustainable alternative jet fuels.

The nation has often counted upon the skills of the aerospace industry to lead the way in technical innovation. Renewable jet fuels offer the opportunity to team aerospace science and technology efforts with those of agriculture, energy, and environment to address the challenges that we face.

Madam Chair and members of the Subcommittee, thank you again for the opportunity to testify on how the aviation community is leading the way to develop and realize the potential of emerging aviation sustainable jet fuels. This completes my prepared remarks. I welcome any questions that you may have.

Senator CANTWELL. Thank you, Dr. Maurice.

Mr. Yonkers, welcome to the Committee. Thank you for being here today.

STATEMENT OF TERRY YONKERS, ASSISTANT SECRETARY FOR INSTALLATIONS, ENVIRONMENT AND LOGISTICS, UNITED STATES AIR FORCE

Mr. YONKERS. Good morning.

Chairwoman Cantwell, Senator Thune, distinguished members of the Committee, it really is a pleasure to be here today, and I thank you for the invitation.

Before I get started, I want to, I would certainly be remiss if I didn't thank you all again for your tremendous support for our Air Force and our airmen that are serving across the globe in many different places every day in the interest of this Nation.

From aviation operations to installation infrastructure both here and abroad, energy enables our core competencies of global vigilance, global reach, and global power, which we need to fly, fight and win. And while the military forces will always be dependent on energy, it certainly is in our best interest to reduce the risk to national security associated with our current energy posture. For the Air Force, this means having access to reliable supplies of energy, and the ability to protect and deliver sufficient fuel to meet our operational, as well as, our training needs.

From an aviation perspective, this includes both increasing our fuel diversity and reducing our demand by becoming more efficient and more responsible in the way we use fuel. As part of our effort to diversify our sources of jet fuel, back in 2006 we began testing and certifying our fleet to use these alternative aviation fuel blends, beginning with a 50-50 blend of traditional JP–8 and synthetic aviation fuel.

We have since expanded that initiative to certify our fleet on a blend of JP–8 and biofuels, and are just beginning to evaluate a third pathway and a fuel blend on alcohol-to-jet.

Last year the Air Force used nearly 2.5 billion gallons of jet fuel at a cost of about $7 billion. We recognize that a $1 change in the price of a gallon results in about $2.5 billion of increased cost, and that has to get paid for in the year of execution. So, as we look at this price volatility, they're not always planned, unless we can look down that road far enough. And we saw this just recently this year with the cost of fuel going from about $3 to almost $4 a gallon.

I want to emphasize that, while we're certainly concerned about cost stability and looking at cost, the primary reason for the Air Force launching into the certification process is to have access to that supply and those supply options that we're going to need to accomplish our mission, no matter where we are across the world.

Our goal is to be prepared to purchase 50 percent of the aviation fuel we use in the United States as alternative aviation fuel blends by 2016. This means that of the 100—of the 1.25 billion gallons we consume in the U.S. on an annual basis, we're looking to purchase about 600 million gallons as alternative aviation fuel blends. However, I will also tell you that our certification process has been so successful that we think we're going to be ready to launch on this well before the year 2016.

The Air Force has the incentive to move swiftly because, in addition to deriving the energy security benefits, we've also found that biofuels burn cleaner and cooler than conventional fossil fuels, and this has a tremendous implication for the wear and tear on our engine parts, and how often we need to put, take engines off of aircraft to recondition them in depots. If we could extend the life, engine life, from 10 or 15 or 20 percent, that would certainly improve our readiness and our ability to go to war.

I also recognize that there are challenges to developing the alternative aviation fuel industry, including the regulatory and economic barriers that have been talked about here today. But, the example that the Air Force supports the goals and intent of the Energy Independence and Security Act, Section 526, it's also key to recognize that all—not all feedstocks comply with this Section 526.

Another potential barrier is the financial commitment to investment in commercial-scale production plants. We'll talk about that today. To reduce the risk to investors, we may need to consider these long-term contracts or other production incentives as a means to attract private capital.

And while we're keeping an eye on all economically-viable alternatives that may be coming to market, the Air Force is also looking at game-changing technologies that will reduce cost, diversify our alternative fuel sources, and optimize our use of domestically produced fuels. And, for example, we're exploring innovative ways to make alternative fuel production units portable and deployable. If we can produce the fuel where we need it, we can reduce the number of convoys we need to use the fuel into our forward operating locations.

Now, we're not just focused on biofuels and synthetic fuels. We're looking at things such as hydrogen as a next generation fuel, starting with a small plant at Hickam Air Force Base in Hawaii with an eye toward expanding this technology to tactical vehicles.

Each of the examples highlight our intent to advance the Air Force's energy security posture, as well as meet our future energy demands, while reducing our greenhouse gas footprint.

Chairwoman Cantwell, Senator Thune, and members of the Committee, that concludes my remarks. I thank you again for inviting me to be here, and I look forward to your questions.

[The prepared statement of Mr. Yonkers follows:]

PREPARED STATEMENT OF TERRY YONKERS, ASSISTANT SECRETARY FOR INSTALLATIONS, ENVIRONMENT AND LOGISTICS, UNITED STATES AIR FORCE

From aviation operations to installation infrastructure within the homeland and abroad, energy enables the dynamic and unique defense capabilities of global vigilance, global reach and global power the Air Force executes to fly, fight and win. . .in air, space and cyberspace. Effective and efficient energy management is not only necessary—it is critical to assuring available energy today and sustainable energy into the future to ensure the Air Force can execute these missions. There is a recognized need to have assured access to reliable energy sources and ensure that sufficient energy is available to meet Air Force operational needs. The Air Force is proud to be a leader in America's ongoing quest for efficient and effective energy use through improved processes, better operational procedures and new technologies, as well as in helping the Nation decrease its dependence on imported oil through alternative fuel and renewable energy usage.

In his recent Blueprint for a Secure Energy Future, President Obama put forward a plan to develop and secure America's energy supplies. At the same time, he challenged Federal agencies to lead by example and help scale up new technologies to support energy security and reduce energy and fuel consumption, resulting in lower costs and reduced pollution. Over the last 4 years, the Air Force has been testing and certifying alternative aviation fuels for unrestricted operational use. The Air Force is certifying its fleet on two fuel blends—the first is a 50/50 blend of traditional JP–8 and synthetic fuel derived through the Fischer-Tropsch process and the second is a 50/50 blend of traditional JP–8 and biomass-derived "hydroprocessed renewable jet" (HRJ). A third fuel blend, a 50/50 blend of traditional JP–8 and alternative fuel derived from cellulosic-based materials, will begin initial feasibility studies within the next few months. The Air Force's alternative aviation fuel initiative is helping the Air Force and the nation improve its energy security posture and is part of the solution to meet some of the President's goals.

The Air Force recognizes that there are many national energy policy objectives, to include the economic impacts of energy costs, the need to reduce greenhouse gas emissions and the national security implications of a high reliance on imported oil. While addressing these challenges, it is of vital importance the Air Force have the energy available necessary to accomplish its missions. Accordingly, the Air Force has developed a comprehensive energy strategy to improve its ability to manage supply and demand in a way that enhances mission capability and readiness. This energy strategy is supportive of DOD's priority program to "Increase Energy Efficiencies" to reduce energy consumption and increase renewable energy.

Air Force Energy Policy: The Air Force's Energy Vision—*Make Energy a Consideration in All We Do*—highlights that energy is central to all aspects of the Air Force's mission execution. In July 2009, the Air Force formally institutionalized its energy program along with its strategy and goals with the issuance of Air Force policy. In December 2009, the Air Force released its Energy Plan, which established "End State Goals" for 2030 and provided a strategic framework to translate formal policy into actionable energy "Focus" areas.

Three primary pillars underpin the Air Force approach's to energy: *Reduce Demand, Increase Supply*, and *Change the Culture*. Each pillar is defined and further developed to include implementing goals, objectives and metrics. This three-pronged approach integrates *demand-side* energy efficiency and mission effectiveness with *supply-side* alternative energy utilization, both of which are enhanced by creating a culture that values energy as a mission-critical resource. The Air Force's alternative aviation fuel program supports the Air Force energy strategy by addressing the need for assured domestic supplies of non-petroleum based aviation fuel.

Program Objectives: The Air Force is motivated by the need to develop a robust, resilient and ready energy security posture, which includes having aviation fuel when and where it is needed to ensure freedom of operation. By increasing the types of fuels available to Air Force aircraft with no degradation in performance, the Air Force is ensuring mission accomplishment and improved national energy security through diversification of supply options. Alternative aviation fuels can have second order effects, multiple fuels sources may insulate the Air Force against volatile oil prices and reduce the environmental impact from aircraft.

The Air Force's long-term goal is to be prepared to cost competitively acquire 50 percent of its domestic aviation fuel requirement via alternative fuel blends by 2016. As part of the goal, the alternative aviation fuel component in the blend will need to be derived from domestic sources and produced in a manner that is more environmentally friendly compared to fuels produced from conventional petroleum. Additionally, any alternative aviation fuel needs to be a drop-in fuel that does not require unique systems or components, or modification to existing systems.

Overview: The Air Force is currently certifying its aircraft and associated support vehicles, equipment and infrastructure for unrestricted operational use on two 50/50 alternative fuel blends. The first blend is a 50–50 mixture of JP–8 and synthetic fuel produced via the Fischer-Tropsch process. The Fischer-Tropsch process starts with a carbon-based feedstock, such as coal, natural gas, biomass or any other carbon-based material, and is gasified before it is converted into a fuel that contains the same chemical properties as traditional petroleum.

The second blend is a 50–50 mixture of JP–8 and HRJ biomass-derived fuel. Under this process, a renewable fuel with properties similar to petroleum is produced from triglycerides, such as plant oils and animal fats. Both the synthetic fuel and the biofuel need to be blended with traditional JP–8, as they do not contain some of the aromatic and other compounds necessary in aviation fuel to safely operate the aircraft.

To ensure an alternative aviation fuel can be used in Air Force aircraft and systems, it undergoes an initial evaluation phase at the Air Force Research Laboratory at Wright-Patterson Air Force Base in Ohio, before undergoing a test and certification phase. This phase, led by the Air Force Alternative Fuel Certification Office, includes engines and flight tests to identify any potential issues with the alternative aviation fuel.

The Air Force is not the only organization evaluating alternative aviation fuels. It is partnering with the airline and aircraft manufacturing industries through the Commercial Aviation Alternative Fuels Initiative to jointly review potential candidate fuels on the basis that the fuels be drop-in with no safety issues or cost increases. The Air Force is also seeking greater efficiencies through joint efforts with the U.S. Navy, the U.S. Army and allied militaries. For example, the Air Force is working with the Canadian Air Force to study of the effects of the HRJ alternative aviation fuel blend on the C–130H aircraft.

Fischer-Tropsch Synthetic Fuel Blend: The Air Force alternative aviation fuel initiative began on September 19, 2006, when a B–52 Stratofortress took off from Edwards Air Force Base in California to conduct a flight test that involved running two of the bomber's engines on a synthetic fuel blend, while the jet's other six engines ran on traditional JP–8 jet fuel. This synthetic fuel blend was a 50–50 blend of traditional JP–8 and Fischer-Tropsch synthetic fuel produced using natural gas as the feedstock.

Following that first flight, the Air Force has achieved a number of successes using a synthetic fuel blend, including the first transcontinental flight, the first supersonic flight, the first aerial refueling and the first fighter demonstration flight. Currently, more than 99 percent of the Air Force fleet is certified for unrestricted operational use of this 50/50 synthetic fuel blend and certification activities are on-track for completion this year. To date, the Air Force has not identified any performance or safety-of-flight anomalies as a result of the synthetic fuel blend, and the military JP–8 Fuel Specification was revised in 2010 to include Fischer-Tropsch synthetic fuel as a blending component. Additionally, the Air Force expects to complete the synthetic fuel certification efforts under budget.

The only remaining Air Force-owned platform left to be certified is the MQ–9 Reaper, which is scheduled to undergo testing and certification later this fall. The only two remaining aircraft in the Air Force fleet requiring certification, the CV–22 Osprey and the F–35 Joint Strike Fighter, are being worked in coordination with the Navy, as both systems are Navy-managed assets.

Bio-mass Derived Alternative Fuel Blend: Following the success of the synthetic fuel certification, the Air Force began evaluation in January 2009 of the 50/50 blend of traditional JP–8 and H RJ biomass-derived fuel. Due to anticipated cost and availability of candidate fuels, the close chemical similarity of HRJ to the previously evaluated synthetic aviation fuel produced using the Fischer-Tropsch process, and the incorporation of "Lessons Learned" from the initial synthetic aviation fuel certification effort, the Air Force determined a fleet-wide certification effort was unnecessary. Rather, the Air Force evaluated only representative aircraft and the most challenging systems from the synthetic fuel certification effort. The remainder of the aircraft will utilize the data obtained during testing of those aircraft and will be certified by similarity.

The Air Force announced its second alternative aviation fuel certification effort when it flew an A–10 Thunderbolt II in March 2010 from Eglin Air Force Base in Florida powered solely by a blend of biomass-derived and conventional JP–8 fuel. This A–10 was the first aircraft ever to be completely powered by such a blend. On February 4, 2011, the Air Force certified the C–17 Globemaster for unrestricted operations using the 50/50 biofuel blend—first Air Force platform certified to fly on the biofuel blend. Only a few months ago, the Air Force's Thunderbirds became the first Department of Defense aerial demonstration team to fly on an alternative aviation fuel blend when three of the six aircraft conducted aerial maneuvers using the biofuel blend at the Joint Service Open House air show at Joint Base Andrews in Maryland. Since the second certification effort began, the Air Force has tested and certified the F–15, C–17 and F–16 aircraft for unrestricted operations, and has demonstrated performance of the A–10 and F–22 using a 50/50 blend of traditional JP–8 and HRJ-derived biofuel. Fleet-wide certification is on track for completion by 2013.

The Air Force has acquired three HRJ fuels in support of its certification efforts, including 200,000 gallons of fuel derived from camelina oil, 200,000 gallons derived from animal fats, and 40,000 gallons derived from waste greases. These fuels were developed domestically, providing an opportunity for U.S. job growth in an industry that improves the Nation's energy security posture. For example, the fuel used to power the Thunderbirds was developed from camelina grown in Montana, while the camelina seed oil used in the H RJ process was cultivated in Montana and Washington State. In both cases, the camelina was grown in rotation with non-irrigated wheat when those fields would otherwise lie fallow, and uses the same infrastructure used for planting and harvesting. The oil was then shipped to Texas, where it went through the refining process to prepare it for use by the F–16s that were part of the Thunderbirds squadron.

To ensure both the synthetic fuel and the biofuel met the Air Force's drop-in requirement, the Air Force tested a C–17 Globemaster on blends of JP–8, Fischer-Tropsch synthetic fuel, and H RJ fuel in August 2010 at Edwards Air Force Base. The tests demonstrated the Air Force could treat both blends as JP–8 drop-ins, as well as co-mingle both alternative fuels. On July 1, 2011, ASTM International, a standards board for materials and products, approved the commercial standard for the renewable fuel which is made from natural plant oils and animal fats and is referred to as "hydroprocessed esters and fatty acids." This approval provides commercially-derived aircraft, including several Air Force aircraft, the option to use 50/50 HRJ blends in their day-to-day operations and provide industry with another potential customer.

Way Forward: Even after certification of the synthetic fuel and biofuel blends is completed, the Air Force will continue to review and evaluate potential alternative aviation fuel candidates. The Secretary of the Air Force recently approved an effort to conduct an initial feasibility demonstration, analysis, and evaluation of the alcohol-to-jet pathway, which uses cellulosic-based materials, such as agricultural and forest waste, to develop an alternative aviation fuel. This initial phase will require no additional funding beyond what has already been provided and enables the Air Force to ensure commercially approved fuels do not compromise the safety and effectiveness of the Air Force systems that may eventually use them. Following the initial study, the Air Force will re-evaluate the alcohol-to-jet pathway to determine if full fleet certification is required. The alcohol-to-jet pathway has been identified by industry as having more commercial potential when compared to both the synthetic and the biofuel blends.

The Air Force has certified nearly all its aircraft and equipment for unrestricted operational use of a 50/50 synthetic fuel blend and is well on its way to certifying its fleet to use a 50/50 biofuel blend. Following full certification, the Air Force will be looking to private industry to develop alternative aviation fuels in commercial-scale quantities. From a feedstock and process perspective, the Air Force is agnostic—as long as the fuel meets the desired performance, environmental and safety specifications, the Air Force will include it in its aviation fuel portfolio. In addition to certifying that all new fuels are safe and effective, the Air Force will use only fuels that comply with all applicable laws and regulations. Even after certification has been completed, the Air Force will not be a producer of alternative aviation fuel, but will use what the market cost competitively provides. This is another reason for pursuing multiple alternative aviation fuel certifications, as it provides the opportunity for the Air Force to ensure it can use any aviation fuel that is commercially available.

Summary: Energy availability and security impacts all Air Force missions, operations, and organizations. The Air Force must have assured energy access to meet the demands of contingency operations abroad and protect the homeland from emerging threats. To enhance energy security, the Air Force is developing a portfolio of renewable and alternative energy sources, including drop-in alternative aviation fuels. By reducing energy demand, increasing the amount and diversity of energy supply, and changing the culture to make energy a consideration in every activity, the Air Force will increase warfighting capabilities, enhance mission effectiveness through efficiency, and help the nation to reduce its dependence on imported oil.

Senator CANTWELL. Thank you, Mr. Yonkers.

Next is Mr. Glover. Thank you very much for being here.

STATEMENT OF BILLY M. GLOVER, VICE PRESIDENT, ENVIRONMENTAL STRATEGY AND AVIATION POLICY, BOEING COMMERCIAL AIRPLANES

Mr. GLOVER. Madam Chair, and Ranking Member, thank you for the opportunity to testify today.

Two years ago I appeared before Congress to testify about the promise of sustainable fuels for the aviation industry. At that time, I stated that Boeing was bullish on sustainable aviation fuels because of the potential environmental benefits, potential economic benefits, and national security implications. What a difference a couple of years make. Today, I'm here to talk about the reality of sustainable aviation fuels.

On July 1, ASTM International approved commercial use of renewable jet fuels derived from natural plant oils and animal fat. ASTM gave the green light for up to 50 percent blend of hydroprocessed fuels, also known as "hydrotreated renewable jet.

"Commercial airlines are already flying on renewable blends. For example, KLM and Lufthansa have already started commercial flights.

The ASTM's adoption of the standard for so-called, "HRJ" reflects an aviation industry cooperative effort that would not have happened as soon as it did without the combined work of the U.S. Air Force, the Federal Aviation Administration, the commercial aviation industry, and many others.

What we learned from our cooperative approach is that these fuels match or exceed the performance of conventional jet fuels. For example, HRJ has excellent thermal stability properties, which may reduce maintenance costs, and very high energy density compared to conventional jet fuels. Higher energy density translates into less fuel per passenger mile.

Nevertheless, while the industry is rightfully pleased with the accomplishments thus far, much work needs to be done to make these fuels commercially available on a widespread basis, and economically competitive. There are more than 20 U.S. biofuel projects in various stages of development, several of which have the potential to produce aviation jet fuel. These projects cover a wide range of feedstocks and process technologies, but all have one thing in common—the need for additional support for near-term development.

Funding for biofuel production has been slowed by the troubled economy and the perception of risks associated with emerging technologies. Just as with the development of the Internet, rural electrification, technical advances growing from the space program, a strong governmental role is essential to assist the sustainable aviation fuels industry through its embryonic development.

There are a number of actions the Government could be doing to spur the production of sustainable aviation fuels—for example, adoption of legislation to allow the U.S. military to enter into long-term contracts for the purchase of sustainable aviation fuels. Providing DoD with the authority to enter into contracts of 10 to 15 years would assist producers in obtaining necessary private financing. Financiers are looking for a commitment of at least 10 years by a party with at triple-A credit rating as a prerequisite for underwriting.

Second, legislation to extend the tax credit under Section 40A of the Internal Revenue Code for producers of biodiesel, renewable diesel, and certain aviation fuels derived from biomass.

And third, funding for research and development on the next generation of sustainable aviation fuels. This is an area where, as Senator Warner mentioned, NASA could be very helpful.

These recommendations were highlighted in the recent report of the Sustainable Aviation Fuels Northwest stakeholder study.

Madam Chair, you've been a very strong supporter, and we greatly appreciate your efforts.

No discussion of incentives for the production of sustainable aviation fuels would be complete, however, without mentioning the programs administered by the United States Department of Agriculture. The 2008 Farm Bill provides a number of important programs aimed at encouraging the production of sustainable biofuels, and while we recognize and understand the issues concerning the budget and the Federal deficit are paramount, we would hope that Congress would find a way forward to continue to support these important programs.

Madam Chair, that concludes my prepared testimony. A complete version has been submitted for the record, and I'm happy to answer any questions.

[The prepared statement of Mr. Glover follows:]

PREPARED STATEMENT OF BILLY M. GLOVER, VICE PRESIDENT, ENVIRONMENTAL STRATEGY AND AVIATION POLICY, BOEING COMMERCIAL AIRPLANES

Madam Chair and Ranking Member:

Thank you for the opportunity to testify today on sustainable aviation fuels.

As you know, The Boeing Company ("Boeing") designs and manufactures commercial and military aircraft, helicopters, missiles, satellites and related components and equipment. We employ approximately 160,000 workers in the United States, and several thousand more overseas.

Introduction

Two years ago, I appeared before Congress to testify about the *promise* of sustainable fuels for the aviation industry. Boeing and four of its airline customers had just completed test flights demonstrating that plant-derived oils could be operated in commercial aircraft without modification to the aircraft or engines. At that time, I stated that Boeing was "bullish" on sustainable aviation fuels because of the potential environmental benefits they could provide relative to reduced life cycle greenhouse (GHG) emissions, the potential economic benefits associated with increased fuel availability, and the national security implications that come with reliance on imported liquid petroleum fuels.

What a difference a couple of years make. Today, I am here to testify to the *reality* of sustainable aviation fuels. On July 1, 2011, ASTM International (formerly the American Society of Testing and Materials) approved the commercial use of renewable jet fuels derived from natural plant oils and animal fat. In an amendment to its D7566 jet fuel specification, ASTM gave the green light for up to a 50 percent blend of hydroprocessed fatty acid esters and free fatty acid (HEFA) fuels—also known as hydrotreated renewable jet (HRJ) fuels—to be mixed with conventional kerosene.[1] Commercial airlines are already flying on blends of HRJ fuels. KLM (flying a Boeing 737–800 aircraft) recently flew the first-ever commercial passenger flight (from Amsterdam to Paris) on a blend of HRJ and conventional jet fuel. Lufthansa recently started daily HRJ-powered commercial flights from Hamburg to Frankfurt. KLM plans to start regular commercial flights later this fall. TUI/Thomson Airways is making its first HRJ-powered commercial flight today, July 28—flying from the UK to Spain with a Boeing 757. In addition, Aeromexico next week will fly a Boeing 777 airplane from Mexico City to Madrid, Spain—thus beginning transatlantic bio-powered service. All these airlines are using a blend of HRJ and kerosene.

The ASTM's adoption of the D7566 standard for HRJ reflects an aviation industry co-operative effort—that would not have happened as soon as it did—without the combined work of the U.S. Air Force (USAF), the Federal Aviation Administration (FAA), and the commercial aviation industry (airlines, as well as aircraft and engine manufacturers and their suppliers). By working together, we were able to perform fuel property tests, materials compatibility testing and engine tests before our first demonstration flights. Once airborne, we were able to put these new fuels through their paces with climbs, engine accelerations and decelerations, windmill engine restarts, starter assisted restarts, and simulated go-around maneuvers. What we learned is that these fuels match or exceed the performance of conventional jet fuel. For example, HRJ has excellent thermal stability properties which may reduce maintenance costs and very high energy density compared to conventional jet fuel. Higher energy density translates to burning less fuel per passenger mile.[2]

[1] The previous D7566 standard, approved by ASTM in 2009, allowed for the use of fuel produced from coal, natural gas or biomass using the Fischer-Tropsch process. Both of the alternative aviation fuels approved by ASTM are complete drop-in substitutes for the petroleum-based fuels currently used in aviation, and are able to use existing fuel transportation and storage infrastructure. The generic term "kerosene" is used in this document to refer to jet fuel derived from petroleum.

[2] The common industry fuel approval process, as embodied in the ASTM process, takes time but results in a very thorough outcome. The thoroughness assures that conforming fuels can

Nevertheless, while the industry is rightfully pleased with its accomplishments thus far, much work needs to be done to make these alternative fuels commercially available and economically competitive.

Sustainable Aviation Fuel Plays an Important Role in the Commercial Aviation Industry's Environmental Commitments

We recognize that the aviation sector, as a key contributor to global GDP, must continually strive to lessen its environmental impact in line with industry growth. To be effective these improvements must be made on a global basis. Over the next 20 years, we expect the global aircraft fleet to more than double, from the current fleet of 17,000 airplanes to more than 35,000. This rapid growth not only presents economic opportunity, but also environmental concerns if that growth is not offset by emission reductions.

It is for this reason that the commercial aviation industry, through work with the International Civil Aviation Organization (ICAO)—the United Nations body that governs all aspects of commercial aviation—has committed to carbon-neutral growth from 2020 and aspires to a 50 percent net reduction in aircraft emissions by 2050 (relative to a 2005 baseline).

To get there, the commercial aviation industry has developed a three-part strategy that we call "planes, practices and fuels." It involves:

- Technology innovation—manufacturers continuing to make more fuel efficient planes through weight reduction programs, aerodynamic improvements and other measures;
- NextGen—accelerating the implementation of advanced air traffic management practices that reduce delays and allow aircraft to fly shorter and more efficient routes; and
- Developing and promoting the commercialization of sustainable aviation fuel as an alternative to conventional jet fuel.

The Boeing 787 Dreamliner and 747–8 are great examples of the industry's technology innovation; each will increase fuel efficiency over predecessor aircraft by approximately 20 and 16 percent, respectively.[3] At Boeing, our strategy is to lead the way in pioneering new technologies for environmentally progressive products and services, and these two aircraft are examples of that effort.

While full build-out and implementation of NextGen will also be a key contributor to reducing aircraft emissions by 12 to 15 percent, neither NextGen nor greater innovative technology will get the commercial aviation industry to a 50 percent reduction in emissions by 2050. That is where sustainable aviation fuels come in—they are so to speak, where the "rubber hits the runway." Sustainable aviation fuel is our industry's sole alternative energy source for the foreseeable future. Unlike other transport sectors, airplanes cannot use plug-in electricity or hybrid power systems.

Specifically, with regard to developing the commercialization of sustainable aviation fuel, Boeing has taken action because we see it as an enabler of greater growth in the commercial aviation industry and therefore in our long term business interest. Our strategy is not however aimed at becoming a fuel producer. We believe that our interest, and frankly the public interest, is better served if Boeing's unique expertise and position in the aerospace industry is focused on accelerating the broad availability of sustainable aviation fuel. That means not just one supply chain success, not just one feedstock success, not just one processing method success—instead it means enabling multiple successes to drive broad commercial availability around the world.

At Boeing, our focus is on sustainable alternatives that have the potential to provide greatly reduced lifecycle greenhouse gas emissions and greater economic benefits associated with increased fuel availability. By sustainable fuels we mean those that comply with robust criteria to ensure that they have significantly better life cycle emissions than traditional fuels, and do not adversely impact food supply, ecosystems, or communities.

It is important to recognize that no one feedstock or processing method will supply all of the aviation industry's needs. Instead, a variety of feedstocks and processing methods will be necessary and they will need to be diversified based upon

---

be used across the existing fleet without modification or further regulatory action. In short, all airplanes are already approved for any conforming fuel. The process can be slowed or accelerated depending upon the availability of test data. If resources like those of the USAF or NASA are readily available, the necessary test data will be available and the process will move faster. Congressional attention toward assuring availability of test resources would be welcome.

[3] Today's jet planes are 70 percent more fuel efficient, which means they produce 70 percent fewer emissions than aircraft produced a mere 50 years ago.

what is commercially available in the locality where the fuel is being produced.[4] It is for this reason that we are participating in a broad range of projects around the world. The central goal of these projects is to develop the scientific, economic and environmental information necessary to develop sustainable aviation fuel resources. (A summary of those projects is attached to my testimony.)

A Blueprint for the Commercial Viability of Sustainable Aviation Fuels

There are more than 20 U.S. renewable fuels projects in various stages of development, several of which have the potential to produce sustainable aviation fuel. These projects cover a wide range of feedstocks and process technologies, but all have one thing in common—the need for additional support for near-term development. Funding for production has been slowed by the troubled economy and the perception of risk associated with investing in emerging technologies. Just as with the development of the Internet, rural electrification, and technological advances growing from the space program, a strong governmental role is essential in assisting the sustainable aviation fuels industry through its embryonic development. Obtaining safe, reliable and environmentally preferred aviation fuels sustains not only the aviation industry, but also builds new agricultural and fuel processing economies as well, all the while providing an important national security hedge against political instability in oil producing regions.

There are a number of actions that the government could be doing to spur the production of sustainable aviation fuels. Of particular importance, Boeing encourages the adoption of:

- Legislation (S. 1079) to allow the Department of Defense (DOD) and branches of the U.S. military to enter into long-term contracts for the purchase of sustainable aviation fuels. Current law does not provide attractive conditions for private investment into production facilities. Providing DOD with the authority to enter into longer term contracts of 10–15 years would assist producers in obtaining necessary private financing. Financiers are looking for a commitment of at least 10 years by a party with a AAA credit rating as a prerequisite for underwriting;
- Legislation to extend the tax credit under Section 40A of the Internal Revenue Code for producers of biodiesel, renewable diesel and certain aviation fuels derived from biomass; and
- Funding for research and development on the next generation of sustainable aviation fuels. Boeing has already begun work with the FAA, the USAF and other industry partners on ASTM approval of new technology pathways to make a bio-derived jet fuel. One of the most promising technologies is the conversion of alcohols to jet fuel. Alcohol-to-jet production processes can work with the existing ethanol and conventional chemical and petroleum production facilities to covert these fuels into aviation fuel. At last count, there were over 150 ethanol facilities in the United States, and for a small capital investment (compared to a new facility), once the fuel is approved by ASTM, they can convert some of the ethanol into aviation fuel.

Madam Chair, it should be no surprise if these recommendations sound familiar; you have been a strong supporter of these legislative initiatives, and we greatly appreciate your efforts. I would also note that these recommendations come directly out of the Sustainable Aviation Fuels Northwest (SAFN) report on sustainable aviation fuels in the Pacific Northwest.[5] SAFN is the Nation's first stakeholder effort to explore opportunities and challenges surrounding the production of sustainable aviation fuels. The report reflects more than 10 months of work and the perspectives of more than 40 stakeholders, and is just one example of the projects that Boeing is involved in around the world.

No discussion of incentives for the production of sustainable aviation fuels would be complete without mentioning programs administered by the United States De-

[4] In the Pacific Northwest, for example, we have identified oilseed crops, algae, municipal solid waste and woody biomass from forest waste as potential sources for the development and production of sustainable aviation fuel.

[5] SAFN was convened by regional leaders in the aviation industry, including Boeing, Alaska Airlines, the operators of the region's three largest airports—Port of Seattle, Port of Portland and Spokane International Airport—and Washington State University, a leader in sustainable fuel research. The regional energy nonprofit, Climate Solutions was retained to facilitate and prepare the report. Full report available at *www.safnw.com*.

partment of Agriculture (USDA).[6] The 2008 Farm Bill provides a number of important programs aimed at encouraging the production of biofuels. For example:

- The Biorefinery Assistance Program (Section 9003)—Provides loan guarantees for the construction or retrofitting of rural biorefineries to assist in the development of new and emerging technologies for the development of advanced biofuels;
- The Biomass Crop Assistance Program (Section 9011)—Provides eligible farmers with matching payments for the sale and delivery of energy crops to biomass conversion facilities;
- Crop Insurance Coverage for Energy Crops (Section 12023)—Requires the Risk Management Agency to develop policies to ensure dedicated energy crops in the same manner as crops used for food and fiber.
- The Bioenergy Program for Advanced Biofuels (Section 9005)—allows the Secretary of USDA to provide production payments to advanced biofuel producers to support expansion of advanced biofuels.

These are just a few of the programs administered by USDA aimed at spurring the growth of energy crops and the production of sustainable fuels. And while we recognize and understand that issues concerning the budget and Federal deficit are paramount, we would hope that Congress would find a way to continue to support these important programs as they apply to advanced biofuels during reauthorization of next year's Farm Bill.

Conclusion

Madam Chair, this concludes my prepared testimony. I am happy to answer any questions you may have.

Regional Solutions: Global Success

Boeing has initiated and participated in a wide variety of projects around the world. Participating in such a wide variety of projects gives us the opportunity to engage with stakeholders, gain perspective, develop scientific, economic and environmental data, and encourage practical steps forward. A summary of projects includes:

- Algal Biomass Organization—Boeing is a founding member of the Algal Biomass Organization, a trade association for algae-to-energy companies and initiatives. Boeing serves on the board of directors and also participates in regular technical and policy projects.
- Commercial Aviation Alternative Fuels Initiative—This partnership among the Air Transport Association, Federal Aviation Administration, Airports Council International-North America and the Aerospace Industries Association explores new U.S. opportunities for sourcing fossil and bio-derived fuels. Boeing initiated the public meeting that led to the formation of CAAFI and participates in technical, research and policy teams.
- "Farm to Fly"—The U.S. Department of Agriculture, Boeing, and the Air Transport Association collaborate to promote development of renewable fuels for aviation. Based on a working together resolution, several key policy recommendations have been proposed to the U.S. Government, with follow-on activities currently underway.
- Latin America Jatropha Sustainability Study—Yale University received funding from Boeing to do the first sustainability assessment of jatropha, a plant suitable for use as an aviation fuel. The peer-reviewed results, based on field data from actual jatropha farms, were released in March 2011.
- Sustainable Aviation Biofuel Evaluation Study—Boeing and PetroChina are leading a comprehensive evaluation for establishing a sustainable aviation biofuels industry in China including agronomy, energy inputs and outputs, lifecycle emissions, infrastructure and government policy support. Other U.S. participants include Honeywell's UOP and United Technologies Corporation,

---

[6] In July 2010, Boeing, the Air Transport Association (ATA) and USDA signed a resolution memorializing their commitment to work together on a "Farm to Fly" initiative to accelerate the availability of a commercially viable sustainable aviation biofuel industry in the United States. The "Farm to Fly" effort has been a very productive forum creating a better understanding of industry potential as well as understanding of how the existing USDA authority can be used to enhance development and use of energy crops to create fuel and jobs. A summary report is being finalized for publication.

while Chinese participants include the Civil Aviation Authority of China, the State Forestry Administration and Air China.

- Sustainable Aviation Fuels Northwest—Sustainable Aviation Fuels Northwest is sponsored by Alaska Airlines, Boeing, the Port of Seattle, the Port of Portland, Spokane International Airport and Washington State University. Boeing initiated and co-funded the project, which convened a diverse stakeholder group looking at the feasibility of developing regionally sourced, sustainable aviation fuels in a four-state region. A final report released in May 2011 is available at *http://www.safnw.com/*.
- Sustainable Aviation Fuels Roadmap—The Sustainable Aviation Fuels Road Map project was developed in collaboration with the Australasian section of the Sustainable Aviation Fuel Users Group (Air New Zealand, Boeing, Qantas, and Virgin Blue) and the Australian Defence Science and Technology Organisation. Boeing initiated and co-funded the regional project, which looked at all phases of developing a sustainable biofuel industry and was coordinated by The Commonwealth Scientific and Industrial Research Organisation. A report issued in May 2011 can be found at: *http://www.csiro.au/files/files/p10rv.pdf*.
- Sustainable Aviation Fuel Users Group (SAFUG)—SAFUG is a global airline coalition accounting for approximately 25 percent of annual commercial aviation fuel consumption. Its members are driving the development of commercial supply chains and supporting the implementation of sustainability standards via the Roundtable on Sustainable Biofuels' global multi-stakeholder processes. Boeing is a founding affiliate and helps coordinate global activity. More information is available at *www.safug.org*.
- Plan de Vuelo—A multi-stakeholder process in Mexico led by SAFUG member Aeropuertos y Servicios Auxiliares (ASA). As part of Mexico's Inter-Ministerial Biofuel Development Commission, ASA is guiding the creation of a Mexican biofuels industry, compliant with global sustainability standards. Boeing worked closely with the Mexican government to facilitate this process and the project report will be released during summer 2011.
- Sustainable Biomass Consortium—Boeing and the École Polytechnique Fédérale de Lausanne created the Sustainable Biomass Consortium, a research initiative for increasing harmonization between voluntary standards and regulatory requirements for biomass for jet fuel. The Consortium aims to lower sustainability certification costs collaborate with civil society and governments on research help, align regional and regulatory requirements, and independently verify the sustainability and traceability of biomass sources.
- Sustainable Bioenergy Research Center—Boeing, the Masdar Institute, Etihad Airways and Honeywell's UOP have established a research institution and demonstration project in Abu Dhabi devoted to sustainable energy solutions. The Sustainable Bioenergy Research Project uses integrated saltwater agricultural systems to develop and commercialize aviation biofuel sources and co-products. Saltwater is used to create an aquaculture-based seafood farming system in parallel with the growth of mangroves and salicornia, a species of saltwater-tolerant plants that offers potential as a sustainable biofuel feedstock.

Senator CANTWELL. Thank you, Mr. Glover. And thank you for being here 2 years ago and pioneering on. We appreciate it.

Mr., is it Todaro? Mr. Todaro, welcome. Thank you very much for being here.

STATEMENT OF TOM TODARO, CEO, ALTAIR FUELS, LLC

Mr. TODARO. Thank you for having me.

So, my name is Tom Todaro. I'm the CEO of a company called AltAir Fuels, based in Seattle, Washington. This is a company that evolved out of an agricultural biotechnology company that we've been running for almost a decade.

If you understand that 90 percent of the cost of a gallon of gasoline, whether it's renewable jet fuel, whether it's renewable is the cost of the feedstock going in. Ours is a company that works with farmers in technology to improve yields to more crops and crop locations into areas that are not environmentally sensitive, and allows probably the most realistic reduction in the cost of fuel over a reasonable period of time.

So, we work today on a variety of crops, but the one that's most widely known is one that's called camelina, which is really just the blue-collar relative of canola. It's been bred to take much less water, much less nitrogen and fertilizer. We can grow it in fallow rotation with wheat, so the farmers can receive a little extra income on land that was otherwise going to be fallowed. We grow them in Washington State, the Dakotas, and other parts of the West Coast.

One of the things that was so important to us was that the life cycle analysis was complete and peer-reviewed from, you know, farm to fly. We've got about an 80 percent reduction in greenhouse gasses related to the fuels we produce, meaning you can fly almost five miles on an airplane powered by fuel we create and have equivalency to emissions as one mile on fuels flown today.

Today, we've produced over 500,000 gallons of this fuel. We've done several dozen military certification flights with the Air Force, with the U.S. Navy, and with the U.S. Army. We've done quite substantial engine testing and commercial flight testing as well. And we're involved, obviously, in the certification of these fuels. You know, we powered an F–18. We broke the sound barrier.

These fuels are molecularly comprehensibly quality tested and can be used in all warfighter and domestic applications for commercial aviation.

The things that we need as a producer—so, we make jet fuel. We do it all the way from the farm. We do the genetic work inside the seeds; we contract with farmers; we move this through a traditional farming infrastructure; we crush out the oil; we convert that oil in collaboration with Honeywell into drop-in fuels.

The things that will surprise people is how close we can get to jet parity in terms of pricing, how quickly we can do it. It's simply an issue of scale.

The things that farmers need is to know that there is a customer that's going to buy their crop. First they need to know the crop is going to exist. So they need crop insurance related to bad weather that affects all other crops that are primarily grown in the U.S.

There is a biomass crop assistance program under the U.S. Department of Agriculture. AltAir Fuels was a happy recipient of an announcement related to that just a couple days ago. But, the long-term renewal of this project is the type of thing that will get farmers to farm. It will give them the support and the comfort they need to take land that was otherwise going to be empty, but they still have to get up and, you know, and, run their tractors, and operate the rest of their farm infrastructure. It allows them what they need to do that until they have a couple of years of data that makes them comfortable that the representations customers are making are accurate.

It would be very helpful to reinstate the requirement with DOD under 526—it was mentioned earlier by Mr. Yonkers. If we want to promote renewable aviation fuel, it would be a shame, in my opinion, to produce such fuels that actually had carbon-negative effects on emissions.

The economic model is really important to understand. Every thousand acres you farm gives you a full-time equivalent job. The projects we're looking at in Washington State and the West Coast—one single facility is a 1,000 farm-related jobs; it's a couple hundred construction jobs; and then, obviously, several dozen to staff the refinery.

So, I guess my concluding point is, we're pretty close to being able to get this done, and if we can get through the next couple of years, we'll prove to all of you and the rest of the country that these fuels are sustainable, they're ready, they're scalable, and they're price efficient.

Thank you.

[The prepared statement of Mr. Todaro follows:]

PREPARED STATEMENT OF TOM TODARO, CEO, ALTAIR FUELS, LLC

Thank you for the opportunity to be here today.

My name is Tom Todaro and I am the founder and CEO of AltAir Fuels, a vertically-integrated refiner of renewable aviation and transportation fuels, as well as renewable chemicals. We are headquartered in Seattle, Washington.

Today I will focus on our approach to producing aviation biofuels, the benefits of this fuel to our land, our farmers, our security, and our ability to pay back our debts, and finally I will discuss ways that Congress can help accelerate the commercialization of these fuels.

First, let's talk about our approach. As an integrated refiner, we secure the raw materials, or feedstock, for our fuels from the most sustainable and economic sources. We refine the feedstock at our facilities and we enter into offtake sales agreements for the finished fuels. Our technology allows us to use many different types of sustainable feedstocks.

Today, our chosen feedstock is an oilseed called camelina, a member of the mustard seed family. Camelina grows in rotation with wheat and other grains in the western United States, including the Dakotas, Montana, Washington, Oregon and California.

Because camelina is designed to be grown on fallow land, in rotation with wheat or other grain crops, it does not displace food crops. In fact, camelina can actually improve soil and water quality compared with the chemical treatments that are typically used on fallow land. After we crush the seeds for the oil, the remaining meal can be fed to livestock and dairy. Camelina has relatively low demands for water or other nutrients, so it is a low-cost crop which helps lower the cost of the finished fuel.

Because of these factors, a peer-reviewed lifecycle analysis of the carbon emissions footprint of camelina-based jet fuel shows an 80 percent reduction compared to petroleum fuels.

To date, we have produced more than 500,000 gallons of renewable jet fuel from camelina. Our fuels have powered numerous aircraft and engines, including: Boeing 747s for Japan Airlines and KLM and engines made by Pratt & Whitney, Rolls Royce and GE. Last month, camelina powered two historic flights—the first trans-atlantic passenger flight—a Gulfstream owned by Honeywell and the first trans-atlantic cargo flight—a Boeing 787.

We have also powered numerous military aircraft, including the A10 Warthog; the F–22, F–16 and FA–18, which last spring made history by breaking the sound barrier on camelina-based jet fuel. The military flights were part of a comprehensive test program with the Air Force, Navy and Army. As Secretary of the Navy, Ray Mabus has said, for our men and women serving in Afghanistan and Iraq, producing renewable fuel at scale is not just an environmental or economic imperative, it is an operational imperative, and all too often, a matter of life and death. The Department of Defense's mission, which we are proud to advance, is to protect our men and women in uniform and make us better warfighters.

One reason camelina-based jet fuel can accomplish this mission is that it can be dropped directly into existing infrastructure—our jet engines and diesel generators can't tell the difference between AltAir's renewable jet fuel and petroleum. The recently created specification for renewable jet fuel approved by the ASTM confirms the viability of our drop-in fuel process, (ASTM D7566–11: Specification for Aviation Turbine Fuel Containing Synthesized Hydrocarbons). When we begin production of our fuels next spring, we expect airlines to be flying on renewable jet fuel from airports on the west coast.

Our long-term plan is to develop Renewable Jet Fuel production facilities across the US, using the locally-sourced, sustainable feedstock. Earlier this week, the USDA approved AltAir Fuels project under the Biomass Crop Assistance Program, or BCAP. This crucial 2008 Farm Bill program is designed to spur the growth of sustainable, non-food energy feedstocks across the country. BCAP provides critical feedstock risk mitigation and will help break the chicken-and-egg problem facing companies like mine that are aiming to build biorefineries to produce homegrown American fuel without enough certainty as to the price and availability of feedstock supply to secure financing. We applaud Secretary Vilsack and his staff's tireless efforts in revising BCAP to make the program more efficient and more workable. Their efforts will empower farmers across the country to reduce our dependence on foreign oil and grow our way out of the national debt while creating jobs in rural America.

As I speak, we are engaging with farmers in Washington, California and Montana to immediately enroll up to 50,000 acres of existing cropland that would otherwise be fallowed to produce camelina with annual funding from BCAP.

The economic impact of our model is significant. We estimate the creation of roughly 400 construction jobs and about 50 full time jobs for each biorefinery. Those numbers soar when you include the increased opportunities for farmers, crushers and transportation-related industries need to support the increased production of camelina. We estimate about 1 job for every 1,000 acres planted; at an estimated 5,000,000 potential acres for camelina, about 5,000 new or retained jobs could be created in rural America.

Given our commercial plans, and our vertically-integrated model, we believe that we can, and will, be producing homegrown renewable jet fuel at the same, or possibly slightly below, the current cost of conventional jet fuel derived almost entirely from foreign oil.

Perhaps more importantly, because we are securing ample supplies of feedstock, we can enter into long-term contracts at fixed prices that bring predictability and stability to customers, whether they are commercial airlines or military combat vehicles.

That said, there are a number of challenges facing this nascent industry. I want to spend my remaining time highlighting ways that Congress can help reduce or remove some of these obstacles to widespread commercialization.

First, we need to continue supporting farmers who choose to grow camelina. Funding for BCAP, which is currently at risk, must be reauthorized if we are going to grow our way out of the national debt, break our foreign oil dependence and create jobs in rural America. We also recommend that Congress consider the use of crop insurance for camelina to help overcome resistance from farmers who can't afford to take a risk on a relatively new energy crop.

Second, we need to ensure that EPA makes a clear determination that camelina-based jet and renewable diesel fuels qualify under the existing Renewable Fuel Standard.

Third, we should enable the U.S. military to enter into long-term contracts—15 years or more—for advanced biofuels. This allows investors to more easily fund production facilities because it lowers their risk.

In summary, let me say that renewable aviation fuel is a reality. There are no technological barriers for either the production or use of these domestic, renewable fuels. This homegrown energy is fueling our jetfighters and commercial planes in the U.S. today. And I look forward to working with this Committee, the Congress and the Obama Administration to ensure that camelina-based fuels and other advanced biofuels continue to propel us toward a more prosperous, energy independent future.

Thank you again for the opportunity to be here today. I am happy to answer questions.

Senator CANTWELL. Thank you very much.

And last, Ms. Pinkerton, thank you very much for being here, and we look forward to your comments.

STATEMENT OF SHARON PINKERTON, SENIOR VICE PRESIDENT OF LEGISLATIVE AND REGULATORY POLICY, AIR TRANSPORT ASSOCIATION OF AMERICA, INC. (ATA)

Ms. PINKERTON. Good morning, and thank you for the opportunity to testify on the impact of fuel policy on our industry, as well as our efforts to develop safe and cost-effective alternatives.

For the Nation, a vibrant alternative fuels industry would mean more jobs, greater national security, and cleaner air.

My name is Sharon Pinkerton, and I'm the Senior Vice President of Policy for the Air Transport Association, representing major passengers and cargo airlines in the United States.

It's really hard to identify any other single product that's more important to the airline industry than jet fuel. Jet fuel touches every aspect of commercial aviation, facilitating our work as we connect people and goods in the global economy, efficiently and safely.

The steady rise of jet fuel prices in the last decade, and the unprecedented price volatility in recent years, has had a tremendous negative impact—not just on the airlines and their employees, but also on the customers and the communities they serve throughout the Nation.

And at this point, Madam Chair, I'd like to thank you for your leadership on that issue of oil speculation. You helped us pass provisions to try to control excessive speculation and, just as importantly, holding the administration accountable to continuing to implement position limits and other of those policies.

Fuel is our largest cost center, representing about one third of our operating expenses. Although U.S. airlines consumed 3.1 billion gallons of jet fuel less, in 2010 than in 2000, they spent a staggering $22 billion more for fuel. And the industry's total fuel spend is expected to rise from $39 billion in 2010, last year, to an estimated $53 billion this year. That's a $14 billion year-over-year increase.

As an industry we're doing everything we can to operate as fuel efficiently as possible and to promote homegrown alternatives to petroleum-based jet fuel. We applaud this subcommittee for holding this hearing and focusing on what the Government can do to help us meet these vital objectives.

What are ATA members doing? We believe we have significant successes to report. U.S. airlines have more than doubled fuel efficiency since 1978. Today, we carry more than twice as many passengers and cargo per gallon as we did then. DOT statistics show that on a system-wide basis, U.S. airlines carried 7.3 percent more passengers and cargo in 2009 than we did in 2000, but we reduced fuel burn and emissions by 14 percent over that same period.

These successes directly reflect our success in leveraging every strategy and technology we can to save fuel. ATA's members have invested billions in advanced airframes and engines; in updating existing equipment with fuel-saving enhancements like winglets, better fan blades, and advanced avionics. As a result of these and many other initiatives, U.S. airlines account for only 2 percent of the Nation's greenhouse gas emissions inventory.

We're also advancing with other industry partners, a global framework for further fuel efficiency and greenhouse gas improvements under the International Civil Aviation Organization, or ICAO. While fuel efficiency enhancements are critical to both our business bottom line and our environmental bottom line, so, too, are commercially available and sustainable alternative aviation fuels that can help put pressure on petroleum-based fuel.

As a co-founding and leading member of CAAFI and other initiatives like Farm to Fly, we're helping lead the way to deployment of alternative fuels. ATA has helped lead the successful effort for specifications certifying two new alternative jet fuels—Fischer-Tropsch and hydroprocessed esters and fatty acids, or HEFA fuels, which can now be used in blends of up to 50 percent with traditional jet fuels, allowing producers to convert a variety of feedstocks into jet fuel. There are other conversion technologies that are underway that you've heard about today, such as alcohol-to-jet.

Our vigorous pursuit of alternatives has sent an unmistakable signal to potential fuel producers and investors—U.S. airlines are committed to making alternative jet fuels viable, and will do their part to overcome the obstacles that may stand in the way. Through pre-purchase agreements, ATA members and airlines have agreed with alternative jet fuel producers to support a number of specific projects, including the production of jet fuel derived from camelina oils or comparable feedstock in the Pacific Northwest, and from agricultural waste and other biomass in California. Such agreements are enabling new business development and jobs across the country.

Despite the significant progress and momentum, we still have challenges. Much needs to be done to bring alternative fuels to commercial viability, and the Government has a central role to play. But here I want to focus on five specific recommendations.

First, commercial aviation should be identified as a top priority for alternative transportation fuels. This is because, while other modes of transportation have other options available, aviation is going to be dependent on liquid, high-energy density fuels for the foreseeable future. At the same time, however, high demand, airport operations in each region of the United States offer a unique opportunity for successful deployment of such alternatives.

Second, existing programs that have been effective in supporting development of alternative aviation fuels must be maintained and, if possible, expanded. Now is the time that we have to move to scalable production and distribution of cost-effective alternative aviation fuels.

Recognizing the need for Government support, Congress has enacted critical programs, such as the biofuel production tax credit, the Biorefinery Assistance Program, and the Biomass Crop Assistance Program. We could lose everything we've achieved if these programs are not maintained.

Third, specific financial support should be provided for promising alternative jet fuel projects to get out, to get off the ground. Even a limited Government commitment would jump-start this industry and build the necessary bridge to a future industry that is entirely funded by private capital. To get there, coordinated government support is needed to establish a proof of concept in the near term.

Fourth, we believe that Congress should encourage near-term environmental benefits. Government policy should be technology and feedstock-neutral. Policy should encourage development of fuels that provide near-term emission benefits, even if GHG reductions are more modest than may be expected in the future development of the biofuels industry. In short, the perfect must not be the enemy of the good.

Fifth, and finally, Congress must ensure that the agencies charged with leading on alternative aviation fuels have the tools to do so. This includes further steps to encourage and empower inter-agency coordination. With respect, with specific respect to the U.S. military, the DoD should be authorized to enter into long-term contracts for alternative fuels and renewable energy.

In sum, ATA and its members are taking action on all fronts to secure and shepherd our energy supply to good use as we power the economy and take the U.S. to global markets. We urge Congress to continue to work with us on this issue that's so critical for our Nation.

[The prepared statement of Ms. Pinkerton follows:]

PREPARED STATEMENT OF SHARON PINKERTON, SENIOR VICE PRESIDENT OF LEGISLATIVE AND REGULATORY POLICY, AIR TRANSPORT ASSOCIATION OF AMERICA, INC. (ATA)

Introduction

As jet-fuel touches virtually every aspect of the commercial aviation business, policies affecting jet-fuel are a core concern for the U.S. airline industry. Recognizing that commercial aviation is an essential driver of the U.S. economy, those policies also should be a core concern for our Nation's policymakers. The Air Transport Association of America (ATA) applauds the Subcommittee for holding this hearing today.

The steady rise of jet-fuel prices in the last decade and unprecedented price volatility in more recent years have had a tremendous negative impact, not only on the U.S. airlines and their employees, but also on the customers and communities they serve throughout the Nation. Congressional action to enhance the level and reliability of fuel supplies and the integrity of aviation fuel markets will help meet those challenges.

Alternative-fuels hold the promise of new, homegrown sources of transportation energy. For the nation, a vibrant alternative-fuels industry would mean more jobs, greater national security and cleaner air. For our industry, a reliable new supply of alternative jet-fuels would help moderate the level and volatility of fuel prices and offer the prospect of further reducing our environmental impact. Our armed forces, with whom ATA is strategically allied in the development and deployment of alternative aviation fuels, would derive similar benefits, further enhancing national security. Everyone wins—except the purveyors of foreign oil.

ATA is working to support development and accelerated commercial deployment of "drop-in" alternatives (fuels that can be used without changing infrastructure) that are safe and deliver environmental, economic and operational benefits, such as supply reliability. We co-founded and co-lead the Commercial Aviation Alternative-fuels Initiative® (CAAFI), a diverse coalition of leading aviation stakeholders dedicated to facilitating alternative aviation fuels. We also are working closely with government agencies, for example, in the *Farm to Fly* initiative, to bring available tools to bear to support aviation biofuels. And our member airlines have executed several pre-purchase agreements for alternative jet-fuel that is soon to be produced.

We have made huge strides, but obstacles remain. Government has a key role to play in helping us overcome them. In terms of general policy matters, it is essential that the government adopt energy policies that increase U.S. energy security, reduce greenhouse gas (GHG) and other emissions, and result in more predictable and stable energy supply and prices. In terms of measures directly relevant to development of alternative-fuels, aviation should be considered a top priority. The aviation industry and would-be alternative jet-fuel suppliers are on the cusp of creating a viable alternative jet-fuel industry. But government support is needed in the near team to provide financial bridging and other tools necessary to help us get over the cusp. We are providing detailed recommendations for how the U.S. Government can—quite literally—help us get the alternative aviation fuels industry off the ground and ensure a future where clean, homegrown jet-fuel is available in significant quantities.

Context for Consideration of Policies to Advance Aviation Fuels

Airlines Are Vital to the American Economy

Commercial aviation is a cornerstone of the economy, driving more than 5 percent of U.S. Gross Domestic Product (GDP). Airlines are at the heart of this, ultimately being responsible for nearly 11 million U.S. jobs and some $370 billion in personal earnings. According to the most recent Federal Aviation Administration (FAA) analysis, every 100 airline jobs help support some 388 jobs outside of the airline industry. In 2010, airlines enplaned 720 million passengers and 18 million tons of cargo on more than 10 million flights. In the same year, U.S. exports by air topped $392 billion and accounted for 31 percent of exports by value.

Commercial aviation also is a key driver of innovation and efficiency. As stated by FAA, "the air transport network contributes added efficiency, technological advancement and versatility that enhance the overall quality of life for U.S. residents and the world as a whole." [1] This not only enhances economic productivity but also enables significant environmental benefits; for example, allowing the production of more goods with fewer warehouses and factories. In turn, this means fewer GHG emissions associated with building and maintaining infrastructure.

Fuel Touches Virtually Every Aspect of Commercial Aviation

No matter what issue or challenge we face, airlines never lose sight of their core mission: safety. Our fuels must meet rigorous specifications that ensure safe operation, whether in the icy cold at 30,000 feet or while filling tanks on the ground at airports crowded with activity.

From a purely business perspective, fuel also plays a critical role. Every penny per gallon costs the industry some $175 million annually, depending on levels of flight activity. The average price of jet-fuel paid by U.S. airlines rose from an average of $0.82 per gallon in 2000 to $2.24 per gallon in 2010. The impact of that dramatic increase is reflected in the fact that although U.S. airlines consumed 3.1 billion *fewer* gallons in 2010 than they did in 2000, they nonetheless spent a staggering $22 billion *more* for fuel. Now, in 2011, the U.S. Energy Information Administration (EIA) is projecting Gulf Coast jet-fuel prices to average $3.06 per gallon (or $128.52 per barrel) for all of 2011, leading ATA to project industry fuel expenditures of $53 billion this year. See Figure 1.

Figure 1. Airline Energy Costs Are High and Poised to Rise

$3.00/Gallon Jet Fuel Would Raise U.S. Airlines' 2011 Fuel Bill by ~ $15 Billion

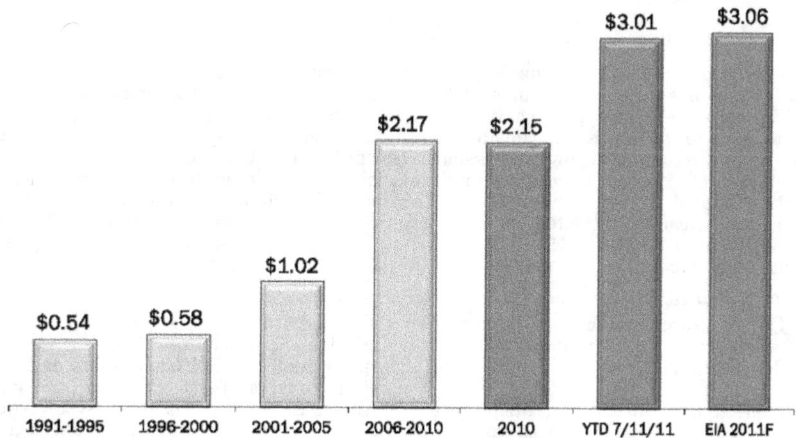

Price level, however, is not the only concern. Especially in recent years, supply disruptions, demand shocks, petroleum futures speculation and other factors have culminated in unprecedented jet-fuel price volatility. See Figure 2. A look at recent

[1] FAA, *The Impact of Civil Aviation on the U.S. Economy* (December 2009) at pp. 6–7.

Gulf Coast prices illustrates the point: From January 2008 through June 2011, monthly average jet-fuel prices ranged from a high of \$3.89 per gallon (in July 2008) to a low of \$1.26 per gallon (in February 2009)—a span of \$2.63 per gallon or 209 percent over just 7 months. On an annualized basis, this difference translates to \$46 billion in airline-industry fuel expenditures, rendering business planning extraordinarily difficult, especially for such a capital-intensive operation. And EIA reported an average price of \$3.14 for the week ending July 15, 2011, the most recent period for which data is available. Among other consequences, the general trend of rapidly rising prices coupled with large, unpredictable price swings has made it increasingly challenging to maintain adequate profitability on a wide number of the routes served by U.S. airlines, resulting in significant scale-backs in seating capacity for many communities and associated job cuts. In the first quarter of 2011, at 33 percent of operating expenses, fuel constituted the industry's top cost; it is estimated to have risen further in the second quarter, although several airlines have yet to report results for that period.

Figure 2. Jet Fuel Volatile Again in 2011

Surpassed \$140/bbl on Rising Crude *and* Refining Crack Spread*

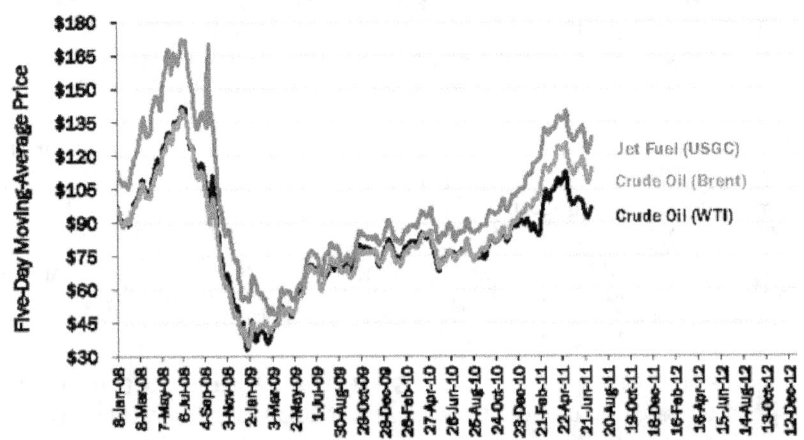

Source: ATA and EIA (for WTI and Brent crude oil and U.S. Gulf Coast jet fuel)

* Refining margin (difference between jet fuel and crude oil price)

Fuel also is central to managing our environmental impact. As detailed below, airlines have a superb environmental record, particularly in reducing carbon dioxide (CO_2) emissions. This performance is closely linked to the financial incentive airlines have to minimize fuel consumption: Lower fuel consumption has the dual benefit of lower costs and lower emissions. The public has benefited from the airlines' relentless efforts to reduce costs and emissions in the deregulated environment, as data compiled by the Department of Transportation show that the average round-trip domestic fare, adjusted for U.S. inflation, was 43 percent lower in 1979 than in 2010 (from \$559 to \$316 in 2010 dollars).[2] Simply put—airlines deliver tremendous economic and environmental bang for the customer's buck.

Commercial Aviation Has a Superb Environmental Record

Our environmental record is particularly strong with respect to the impact most closely related to combustion of jet-fuel: emissions. For example, the latest EPA GHG inventory shows that commercial aviation's domestic GHG emissions declined 18 percent from 1990 to 2009, even though we transport far more cargo and passengers today. Bureau of Transportation Statistics data show that on a systemwide basis, U.S. passenger and cargo airlines carried 14 percent more passengers and cargo in 2009 than in 2000 while reducing fuel burn and emissions by 7.3 percent. Similarly, fuel efficiency (measured by revenue ton miles per gallon) has more than doubled since 1978; stated differently, for every mile flown, today we carry more

[2] ATA analysis of data compiled by the U.S. Department of Transportation in the Origin-Destination Survey, more commonly known as Data Bank 1A.

than twice as many passengers and cargo per gallon than we did in 1978. And EPA GHG inventory shows that commercial aviation's share of GHG emissions in the United States is only 2 percent of the Nation's GHG emissions today.

The U.S. airlines have accomplished this tremendous record by investing billions in new equipment, infrastructure and technology to maximize fuel efficiency. This includes purchasing advanced airframes and engines and updating existing equipment with fuel-saving enhancements like winglets, better fanblades and advanced avionics. We also seek to maximize efficiency of operations in the air by taking advantage of new procedures like continuous descent approach (CDA), required navigation performance procedures (RNAV) and reduced vertical separation minima (RVSM). Other measures maximize fuel efficiency while on the ground, like taxiing on one engine where operationally feasible and utilizing electric gate power instead of our planes' auxiliary power units (APUs) while parked at the gate. Measures as banal as reducing aircraft weight by eliminating unneeded magazines and replacing catering carts with new light-weight carts or washing fan blades more often also can result in small but cumulatively significant fuels savings.

It bears emphasis that implementation of some of these measures are not fully within the control of airlines, but require government action. For example, fuel-saving procedures must be approved by FAA, and broad access to these procedures will depend on full and cost-effective implementation of NextGen, which encompasses the suite of technologies and initiatives required to transform today's antiquated ground-based air traffic navigation and surveillance system into a state-of-the-art satellite-based system. Utilizing this system, FAA and the airlines will be able to route flights more efficiently, precisely and directly, leading to lower fuel consumption and emissions while increasing safety by enhancing situational awareness for pilots and controllers.

And the industry is not stopping with these measures. ATA and its members are part of a worldwide aviation coalition that has put a strong proposal on the table for further addressing aviation CO_2 under the International Civil Aviation Organization (ICAO), the United Nations body charged by treaty with setting standards and recommended practices for international aviation. Our focus is on getting further fuel efficiency and emissions savings through new aircraft technology, sustainable alternative aviation fuels and improvements to air traffic management and infrastructure.

Under our proposal, all airline emissions would be subject to collective emissions targets requiring industry and governments to do their part. The emissions targets include collective industry commitments to:

- Continue the industry's fuel (and, hence, CO_2) efficiency improvements, resulting in an average annual CO_2 efficiency improvement of 1.5 percent per year on a revenue ton mile (RTM) basis through 2020;
- Cap industrywide CO_2 emissions from 2020 (carbon-neutral growth) subject to critical aviation infrastructure and technology advances achieved by the industry and government; and
- Contribute to an industrywide goal of reducing CO_2 emissions by 50 percent by 2050, relative to 2005 levels.

Significantly, at its 2010 Assembly, ICAO adopted much of the industry's framework. While more work is needed to flesh out this framework, the global aviation industry is moving forward with its emissions-savings initiatives.

Airlines Are Uniquely Positioned to Benefit from and to Facilitate the Emergence of Alternative Fuels

While other sectors and modes of transportation can be powered via a variety of energy sources, including electricity, nuclear, solar, hydrogen and wind, to name a few, airlines will be flying aircraft and engines requiring liquid, high energy-density fuels for the foreseeable future. There simply is no realistic prospect for the next several decades that commercial aircraft will be powered by batteries, solar cells, fuel cells, hydrogen or other alternatives. This is primarily a function of the reality that the useful life of aircraft and aircraft engines is very long, and that the pipeline for development of new aeronautics technologies is even longer. As a result, airlines will be flying aircraft designed to operate on fuels that meet the performance characteristics of traditional petroleum-derived jet-fuel for decades to come. Consequently, while other modes and sectors may benefit from the emergence of other energy and fuel alternatives, commercial aviation can benefit only if it has access to significant supplies of liquid alternative-fuels that meet the rigorous safety and performance criteria required of current petroleum-based fuels.

Commercial aviation, however, also offers unique benefits to prospective fuels producers. First, fuel demand is highly concentrated. The 40 largest airports account for an estimated 90 percent of all jet-fuel U.S. demand while the top 10 airports account for about half of demand. The country's largest airports—Los Angeles (LAX), New York-Kennedy (JFK), Chicago O'Hare (ORD) and Atlanta (ATL)—each demand more than one billion gallons of jet-fuel annually. Demand from Air Force bases and Navy installations is also highly concentrated. Thus, airports essentially compose a network of markets that alone could support all the output from alternative-fuels production facilities. In addition, with high-demand nodes across the country, the aviation industry can support production from the full gamut of potential producers, who will rely on different feedstocks, depending on where they intend to operate.

Alternative Jet-fuels Offer a Rare Opportunity

Development of alternative jet-fuels offers a rare opportunity to meet disparate but beneficial objectives. A vibrant alternative jet-fuels industry would create American jobs and spur economic development in areas most hit by the recession. Rural America would benefit greatly from access to new markets for new agricultural biomass crops while industrial areas would be revitalized through construction of new or revitalization of mothballed refinery operations. At the same time, a stable, domestic supply of alternative jet-fuel would improve our Nation's security by reducing our dependence on foreign oil and improve national economic security by improving our trade balance. In turn, stable, homegrown production of alternative jet-fuels will introduce competition to petroleum-based jet-fuels and a moderating force on price levels and volatility. This would be a very welcome change for airlines that have struggled to manage their businesses as prices driving their number-one cost center have steadily risen and fluctuated sharply in recent years. Undoubtedly, the conditions necessary to foster a financially healthy, vibrant and growing commercial aviation sector—so vital to the overall health and vitality of U.S. commerce—would improve, further benefiting the broader economy as airline-driven growth is known to generate numerous jobs beyond the aviation sector.

Sustainable alternative-fuels also will allow our industry to grow while reducing its emissions of GHGs and emissions with local air-quality impacts. Such fuels also could be used in our ground support equipment (GSE), removing costs associated with management of separate fuels and further reducing emissions.

The U.S. military, which has been a very active ATA partner in the pursuit of jet-fuel alternatives, shares many of these same interests. To formalize this working relationship, on March 19, 2010, ATA and the Defense Logistics Agency's Defense Energy Support Center (now known as DLA Energy) signed a "Strategic Alliance for Alternative Aviation Fuels."[3] Like airlines, jet-fuel represents a significant share of costs to the U.S. military, particularly the U.S. Air Force. Rising and volatile prices wreak havoc on military budgets and present significant challenges for military planners, especially as combat logistics become increasingly complex and supply lines extend over often mountainous or desert terrain. At the same time, GHG emissions from military jet operations represent a large portion of the Federal Government's carbon footprint. Access to stable, domestically produced supplies of low-carbon alternative-fuels would allow the armed services to address all of these concerns in the same manner it would enable commercial aviation to address the parallel concerns as discussed above.

The opportunities presented by the prospect of viable alternative jet-fuel are reflected in the four specific requirements we set out as conditions for use:[4]

1. *Safety/Fuel Quality*: To ensure safety, commercial jet-fuel must meet precise technical and operational specifications, and jet engines are designed to work with jet-fuel having these specific characteristics. The fuel must meet regulatory and standards-making organization specifications including, but not limited to, ASTM D1655 and others referenced and required by the FAA.

2. *Environmental Benefit*: We seek alternative-fuels that will meet accepted criteria to be more environmentally friendly than traditional jet-fuel, in particular resulting in a reduced emissions profile on a life-cycle basis, without compromising critical uses of relevant feedstocks.

3. *Supply Reliability*: Alternative jet-fuels must be "drop in" fuels, meaning they must satisfy technical and functional criteria that make them fungible with traditional, petroleum-based jet-fuel and allow them to be commingled within the existing national fuel transport, storage and logistics infrastructure, as well as within individual airport and airline systems.

---

[3] *http://airlines.org/News/Releases/Pages/news_3-19-10.aspx.*

[4] *http://airlines.org/Energy/AlternativeFuels/Pages/CommercialAviationAlternativeFuelsTheATACommitment.aspx.*

4. *Economic Feasibility*: Alternative jet-fuels must be economically feasible from the perspectives of both suppliers and purchasers.

Airlines Have Been Working Diligently to Support Development of Alternative Fuels

ATA and its member airlines are committed to finding safe, environmentally preferred, operationally reliable and economically feasible alternatives to conventional petroleum-based jet fuel. This is no easy task. Realizing the deployment of significant quantities of viable alternative jet fuel will require overcoming significant technical and financial hurdles. To meet this challenge, we are proactively addressing the commercial, environmental and safety issues associated with developing and commercializing promising technologies that can meet our needs.

Five years ago, together with the Federal Aviation Administration (FAA), Airports Council International—North America (ACI–NA) and the Aerospace Industries Association (AIA), we founded CAAFI, a coalition that brings together leaders in the aviation community (including airlines, airframe and engine manufacturers and airports), alternative-fuels providers, universities and government stakeholders to exchange information and work to make alternative aviation fuels a reality.

ATA also has directly engaged government stakeholders. Specific engagement with the U.S. military includes the aforementioned alliance with DLA Energy. On March 30, 2011, in a bellwether speech on America's energy security, President Obama recognized the role of precisely this type of partnership by issuing the following directive:

> . . . our Air Force used an advanced biofuel blend to fly an F–22 Raptor faster than the speed of sound. In fact, the Air Force is aiming to get half of its domestic jet-fuel from alternative sources by 2016. And I'm directing the Navy and the Departments of Energy and Agriculture to work with the private sector to create advanced biofuels that can power not just fighter jets, but trucks and commercial airliners.[5]

ATA also has joined Boeing and USDA in putting together and leading the *Farm to Fly*[6] initiative aimed at advancing a comprehensive sustainable aviation biofuels rural development plan. To achieve this, we have engaged the U.S. agencies with authority to spur development of aviation biofuels (including USDA, DOE, DOT, FAA and DOD) and academia to ensure that Federal programs are aligned and modified to recognize and enhance the eligibility of feedstocks, conversion technologies and supply chains most conducive to the production of aviation biofuel.

The *Sustainable Aviation Fuels Northwest (SAFN)* initiative,[7] led in part by ATA member Alaska Airlines, together with the Port of Seattle, Port of Portland, Spokane International Airport, Boeing and Washington State University, is another example of a coalition effort in which we have been engaged to enable sustainable alternative aviation fuels. More than 40 organizations representing a broad range of stakeholders participated, including aviation, biofuels producers, environmental NGO's, agriculture, forestry, Federal and state government agencies, and academic institutions. This effort culminated in a report that detailed opportunities for and measures needed to foster the development and deployment of alternative jet-fuels derived from sustainable biomass grown in the northwestern United States.

An extremely important step in alternative aviation fuel development is fuel certification. Accordingly, ATA and other stakeholders such as FAA have made great strides in this area. Before the fuel can be approved for commercial use, it must meet rigorous safety and performance standards set out in the applicable specification, which is controlled by ASTM International, an organization devoted to the development and management of standards for a wide range of industrial products and processes. This specification, in turn, is included in FAA product approvals and required air-carrier manuals. The process for securing new specifications for alternative fuels is exacting. Supporting test data is referred to ASTM Subcommittee D02J to review and the sponsors of the new fuel write a new specification for that fuel. The specification and required research reports are then reviewed and voted upon by the technical experts. The specification allowing use of Fischer-Tropsch (FT) in blends of up to 50 percent with traditional jet fuels was approved in September 2009[8] and the specification for hydroprocessed esters and fatty acids (HEFA) jet

[5] *http://www.whitehouse.gov/the-press-office/2011/03/30/remarks-president-americas-energy-security.*

[6] *http://airlines.org/News/Releases/Pages/News_07-21-10.aspx.*

[7] *http://www.safnw.com/.*

[8] *http://www.astmnewsroom.org/default.aspx?pageid=1895.*

fuels (again in up to 50/50 blends) was formally approved just a few weeks ago, on July 1.[9] FT is a chemical process that can convert a variety of feedstocks (including fossil-fuel sources like coal and natural gas, as well as biomass) in liquid fuels. HEFA fuels can come from many regionally grown and processed sustainable feedstocks. Both types of fuels can offer significant GHG reductions relative to conventional jet fuel. The successful conclusion of the specification approval process for these two fuels has paved the way for additional fuels to be examined and approved in the future.

ATA also is working to confirm agreed-upon methodologies for determining the emissions profile of alternative fuels. This can be extremely complicated, requiring close analysis of the emissions associated with each link in the "life cycle" of a fuel, including production of the feedstock, transportation of the feedstock, processing and refining, and transportation of the final product. Significant work has been done in this area, including by agencies implementing alternative-fuels programs like EPA (which evaluates fuels in implementation of the Renewable Fuels Standard established under the *Energy Policy Act of 2005* and expanded under the *Energy Independence and Security Act (EISA) of 2007)* and the California Air Resources Board (CARB, which evaluates fuels under its Low Carbon Fuels Standard—LCFS). While these programs establish basic methods and criteria for life cycle analysis, ATA is working through CAAFI to confirm jet-fuel-specific applications. Other, broader criteria for assessing the "sustainability" of fuels also have been considered by various entities, which address environmental and other impacts beyond emissions. CAAFI also is working to identify issues relevant to alternative jet fuels and reach acceptable resolutions.

The airlines' initiative in tackling these issues with our partners has sent a clear unmistakable signal to potential fuels producers and investors: airlines are committed to making alternative jet-fuels a reality and will do our part to overcome the obstacles that may stand in the way. Scores of companies eager to meet our demand have emerged and are themselves helping to resolve these issues, again largely through participation in CAAFI. The fruits of this labor are apparent in that ATA member airlines have agreed with alternative-fuels producers to support a number of specific projects, including:

1. On December 15, 2009, 15 airlines from the United States, Canada, Germany and Mexico signed memoranda of understanding (MOU) with:

- AltAir Fuels LLC ("AltAir"), involving camelina and potentially other crops in the western United States for the production of 75 million gallons per year, over a 10-year period, of jet-fuel and diesel fuel derived from camelina oils or comparable feedstock, refined in the State of Washington.[10]
- Rentech, Inc. ("Rentech") contemplating the production of approximately 250 million gallons per year of synthetic jet fuel at a facility in Adams County, Miss. ("Natchez Project"). The fuel will be derived from coal or petroleum coke, with the resultant carbon dioxide sequestered. This drop-in synthetic jet-fuel will have lower regulated emissions and a lower carbon footprint than traditional jet fuel. Rentech intends to potentially further reduce the carbon footprint by integrating biomass as a feedstock.[11]

2. On August 18, 2009, eight U.S. airlines—Alaska Airlines, American Airlines, Continental Airlines, Delta Air Lines, Southwest Airlines, United Airlines, UPS Airlines and U.S. Airways—signed an agreement with Rentech and Aircraft Service International Group (ASIG) to purchase up to 1.5 million gallons per year of renewable synthetic diesel for use in ground service equipment at LAX beginning in late 2012 or 2013, with urban woody green waste from the Los Angeles area.[12]

3. On June 20, 2011, a core group of airlines signed letters of intent with Solena Fuels, LLC ("Solena") for a future supply of jet-fuel derived exclusively from biomass to be produced in northern California. Solena's "GreenSky California" biomass-to-liquids (BTL) facility in Northern California (Santa Clara County) will utilize post-recycled urban and agricultural wastes to produce up to 16 million gallons of neat jet fuel (as well as 14 million gallon equivalents of other energy products) per year by 2015 to support airline operations at Oakland (OAK), San Francisco (SFO) and/or San Jose (SJC). The project will divert approximately 550,000 metric tons of waste that otherwise would go to a landfill while pro-

---

[9] *http://www.astmnewsroom.org/default.aspx?pageid=2524.*
[10] *http://airlines.org/News/Releases/Pages/news_12-15-09.aspx.*
[11] *http://airlines.org/News/Releases/Pages/news_12-15-09.aspx.*
[12] *http://airlines.org/News/Releases/Pages/news_8-18-09.aspx.*

ducing jet-fuel with lower emissions of greenhouse gases and local pollutants than petroleum-based fuels.[13]

And more such projects are in the works.

Government Has an Essential Role to Play in the Success of Alternative Fuels

Commercial aviation is doing all that it can to minimize fuel burn, reduce emissions, enhance stability of supply and foster the production of alternatives. But we cannot do it alone. We need sustained leadership and support from the U.S. Congress and administration. We applaud the leadership already provided by the Department of Transportation-commissioned Future of Aviation Advisory Committee (FAAC), which under the direction of Secretary of Transportation Ray LaHood, reached consensus on several recommendations regarding what government needs to do to help ensure the viability and global competitiveness of the U.S. aviation industry,[14] including:

- Accelerate NextGen implementation by providing government financial incentives to airline operators for equipage;
- Expedite the most cost-beneficial elements of NextGen, including ADS–B and performance-based procedures;
- Ensure that the Federal aviation tax burden does not undermine the viability and competitiveness of the airline industry;
- Mitigate jet-fuel price volatility by supporting Federal regulatory efforts to mitigate the impact of speculative activity on the price of oil; and
- Reduce the impact of aviation on the environment through the use of sustainable fuels and improved aircraft technology.

Many of these points are echoed in our recommendations for action, which fall into two categories: (1) general policies affecting energy and fuel and (2) measures directly relevant to development of alternative-fuels.

Recommendations Regarding General Policies Affecting Energy and Fuel

A. *Government must adopt energy policies that increase U.S. energy security, reduce GHG and other emissions, and result in more predictable and stable energy supply and prices.* The enactment of the Dodd-Frank Wall Street Reform Act was an important step toward eliminating speculation-driven price volatility in oil markets. But it is equally important now that the Commodity Futures Trading Commission's implementing regulations to curb speculation that distorts oil markets are consistent and meet the objectives of the law.

B. *Congress must support programs that enable ever-increasing fuel efficiency in the aviation sector.* This includes:

a. *Funding and encouraging a business-case approach to implementation of NextGen.* Cost-effective implementation of NextGen, in addition to many other benefits (including reduction of delays even as capacity of the system is increased) will save fuel and Congress needs to fully support it.

b. *Restoring funding for basic aeronautics research and development at the National Aeronautics and Space Administration (NASA) and FAA.* With the airlines' support, commercial aircraft and engine manufacturers have succeeded in consistently improving the safety, reliability and performance of commercial aircraft. Improvements in fuel efficiency have been accompanied by improvements in noise and emissions. Unfortunately, in the near future, no major breakthrough in either aircraft or engine design is expected because of the enormous effort and cost of engineering research and development. Over the past several years, the Federal Government has significantly reduced funding for FAA and NASA aeronautics research and development programs, which are critical in moving airframe and engine technologies forward. Countering this trend requires the Federal Government to restore and increase funding for aeronautics research.

c. *Supporting a global sectoral approach to regulation of aviation GHG emissions to be overseen by ICAO.* As discussed above, U.S. airlines have joined the global aviation industry in adopting an ambitious set of near-, mid- and long-term targets to further mitigate GHG emissions from our industry under a global sectoral approach. Congress should endorse this approach.

[13] *http://airlines.org/News/Releases/Pages/news_6-20_11.aspx.*

[14] *http://www.dot.gov/faac/docs/faac-final-report-for-web.pdf.*

Recommendations for Measures to Support Development and Deployment of Alternative Fuels

A. *Commercial aviation should be identified as a top priority for alternative transportation fuels.* As previously discussed, while other sectors and modes of transportation have other options available, aviation will be dependent on liquid, high energy-density fuels for the foreseeable future. At the same time, however, with concentrated demand nodes in each region of the United States and an industrywide commitment to ensure that alternative aviation fuels are successful, aviation presents a unique opportunity for successful deployment of such alternatives. We ask the Subcommittee to support policies and initiatives that prioritize alternative-fuels for aviation.

B. *Government law and policy should not discriminate among alternative-fuel technologies.* Commercial aviation will use *any* alternative fuel that meets the four criteria laid out above concerning safety, environmental benefit, supply reliability and economic feasibility. The appropriateness of using certain feedstocks or processes must not be prejudged or disqualified for use based on other agendas.

C. *Congress should encourage near-term environmental benefits.* Policy should encourage development of fuels that provide near-term emissions benefits, even if GHG reductions are more modest than may be expected in the future development of the biofuels industry in the United States. Policies that require fuels to meet elevated emission-reduction targets as a precondition to receiving government support risk erecting unnecessary barriers to achieving greater reductions in the future. In short, the perfect must not be the enemy of the good—especially where "the good" has the potential to mature into "the great."

D. *Government policy must ensure coordination among various government agencies with authority to provide support to alternative-fuels development, including the DOT/FAA, USDA, DOE and DOD.* In our experience, these agencies are doing what they can within their existing authorities and mandates to coordinate activities and leverage mutually reinforcing programs. Congress should take further action to encourage and empower this type of interagency coordination and commingling/aggregation of fiscal and human resources.

E. *To support our military and the development of alternative-fuels, we also ask Congress to authorize DOD to enter long-term (up to 20-year) contracts for alternative-fuels and renewable energy.* To secure investment in capital-intensive alternative-fuel production facilities, providers must be able to demonstrate revenue streams extending out at least 10 years but ideally more on the order of 20 years. Without long-term contracting authority, the military simply will not be able to participate meaningfully in efforts to spur construction of alternative-fuel production capacity. Congress needs to remedy this.

F. *It is critical that existing programs that have been effective in supporting development and deployment of alternative aviation fuels be maintained and, if possible, expanded.* First, it is vital that cellulosic biofuel producer credit be extended. Second, programs direction Federal agencies to help America transition to alternative-fuels need to be funded. These include the Biomass Research and Development Initiative, Biorefinery Assistance Program, Bioenergy Program for Advanced Biofuels, Marketing Assistance Loans and Loan Deficiency Payment Programs, Biomass Crop Assistance Program, Crop Insurance Coverage for Energy Crops, and National Institute of Food and Agriculture.

G. *Many projects with the potential to produce alternative jet-fuels already have been developed and tested but need additional funding for near-term development.* Economic conditions have made credit and investment difficult to come by—it is even more difficult for emerging technologies. In this environment, government support is essential to assist the alternative-jet-fuels industry through this early stage in its development. Marshaling existing funding and other mechanisms across agencies to support one or more projects with the aim of proving production of significant quantities of alternative-fuels is possible will go a long way toward demonstrating commercial viability to reluctant private capital. A limited government commitment would "jump start" this industry and build the necessary bridge to a future in which the industry is entirely funded by private capital. To be clear, ATA is not calling for perpetual government funding. For an industry that is self-sustaining to emerge, however, requires "proof of concept" in the near term and this is where government support is necessary and should be focused.

A final point deserves emphasis: The last thing we need is more taxes on commercial aviation. Also particularly relevant here is the European Union's Emissions Trading Scheme (EU ETS), which imposes a steep tax on jet-fuel consumed by U.S. airlines for flights to or from Europe, even when they are in U.S. airspace, on the ground in the United States or over the high seas. Such taxes are counterproductive—siphoning slim resources from airlines and compromising our ability to make the types of investments in technology that have enabled us to transport more and more people and goods, even as we reduce our environmental impacts. Commercial air transportation already is one of the most heavily taxed businesses in the country, facing rates comparable to those of alcohol and tobacco, which are designed to discourage their consumption. Discouraging air transportation, which drives the global economy with still more taxes is the last thing we should be doing, particularly in these economic times. We urge the Subcommittee to join the administration's opposition of the application of the EU ETS to U.S. airlines, and to oppose new or increased taxes here at home.

Conclusion

We will continue to do everything we can to minimize fuel burn, reduce emissions, enhance stability of supply and foster the production of alternatives. ATA looks forward to working with the Subcommittee to help spur government actions and leadership necessary to realize these objectives.

Senator CANTWELL. Thank you very much, Ms. Pinkerton. And we appreciate your giving us those specific recommendations.

I want to start with questioning.

Mr. Yonkers, you talked about 2016 in, your goal to achieve a 50 percent domestic aviation fuel requirement, and then you alluded to, maybe you could do it even sooner, and it may be rosier than even that projection. Can you talk about what are the milestones to getting to 2016? What you think those challenges are, how you might see us proceeding to achieve that goal.

Mr. YONKERS. Well, Senator Thune talked about the initial Fischer-Tropsch certification, and we're 99 percent of the way there. There are two platforms that we're, you know, still looking to complete that final touch on the Fischer-Tropsch alternative synthetic fuel approach, and that's Global Hawk and Reaper, the remotely-piloted aircraft.

When we turn over and look at the HRJ, we're going to be completed with that certification program probably by the end of 2013. Then as we look down the road at the next pathway, the alcohol-to-jet, probably the 2014–2015 timeframe.

So, essentially what I think we're on a glidepath to achieve here is a full certification for all three pathways by the year 2014 and 2015. As we go down this road, we learn, we get better at what we do. We have a pathfinder approach, so that we're not going and recertifying, or certifying every aircraft platform; that we can by extension and extrapolation look at the family of particular biofuels or synthetic fuels and arrive at the same conclusion, because if they fly on one aircraft without problems they can fly on another.

Senator CANTWELL. And what is the mix? You know, obviously, we hear about synthetic fuels. How green are these synthetic fuels that you're talking about?

Mr. YONKERS. Well, there's certainly a drawback with the synthetic fuels in particular, whether it's coal-derived or whether it comes from natural gas. And in some cases, the coal, I think our results are showing that from a greenhouse gas point of view, it's probably twice of what conventional fuel is. The biofuels, as you've heard from some of the members on the panel today, are upwards to 80 percent better. Our results show that 60 to 70 percent. So, pretty close parity with what we're seeing out in the private sector.

Senator CANTWELL. Ms. Pinkerton, could you talk about the EU and how important it is that we move forward on a green source of synthetic fuel, or, how is the airline industry looking at this from a commercial side?

Ms. PINKERTON. Well, let me address your question about the EU. As I think most of you know, the EU is proposing a unilateral cap-and-trade system on U.S. airlines, starting in U.S. airspace. They are proposing to move forward with this. They've already started collecting data. There was a hearing yesterday on the House side in which the Administration, thankfully, testified in opposition to that unilateral approach.

But, I have to say, alternative fuels are absolutely a critical element of the airlines' ability to meet their global commitments through ICAO. We've committed to 1.5 percent fuel efficiency improvements; to carbon-neutral growth starting in 2020. And we can't get there by ourselves. We're going to need to implement NextGen, but even with NextGen, alternative fuels are clearly a critical part of our ability to meet our international goals.

Senator CANTWELL. What happens if the EU continues to act unilaterally?

Ms. PINKERTON. Well, I think there are a variety of options. Legislation has been introduced in the House side. It was introduced last week—bipartisan legislation to essentially prohibit U.S. airlines from participating in the EU ETS scheme. But that being said, the administration has taken the lead to sit down with the Europeans. They did so in June of this year and, essentially, laid out the U.S. Government's position—and by the way, we don't stand alone. Latin America, the Chinese, essentially the rest of the world is with us in opposing the EU scheme.

We're hopeful that through these diplomatic channels we're going to be able to convince the Europeans to stand down, to work with us in ICAO in achieving a global framework to address climate change, because climate change is a global issue.

But it's unclear. There's potential for litigation in the U.N. process. That's certainly one option. But we hope it doesn't come to that.

Senator CANTWELL. Thank you.

I want to go to Senator Thune, and then I'll come back to the panel.

Senator THUNE. Thank you, Madam Chairman.

Mr. Yonkers, what's the average that the Air Force now pays for a gallon of alternative fuel?

Mr. YONKERS. Right now, Senator, we're paying about $35 for the HRJ-derived fuels. That's down from $65 and $200 for when we started. So, the price is dropping. And, again, keeping in mind that we're buying batches of about 30,000 gallons or so. So, it's not full production by any stretch. On the Fischer-Tropsch side it's much more reasonable—$3 to $4 is the current cost.

Senator THUNE. Does that price-per-gallon take into account the funding for research and development and testing and evaluation?

Mr. YONKERS. Certainly on the supply side it does. The Air Force is also spending about $20 million a year for our Air Force Research Laboratory and others that are involved in the fuel certification process, and some of these other things that I talked about in my opening remarks.

Senator THUNE. What price range per gallon would alternative aviation fuel need to be at so it's affordable for the Air Force to use it in a majority of its fleet?

Mr. YONKERS. Well, certainly lower is better.

Senator THUNE. Yes.

Mr. YONKERS. You know, and that's what's promising about the discussion we're having today, particularly with the alternative fuels and the biofuels, and the alcohol-to-jet, which is even more promising, I think. But I think, you know, our position has always been, we're realistic here so that, you know, we're going to be willing to pay whatever the market price is for jet fuel.

Senator THUNE. Can you explain the importance to national security for commercial aviation and DOD to partner in the development of alternative fuels?

Mr. YONKERS. Well, certainly, when we partner together, you know, our 10 percent added to their 90 percent sends a very strong demand signal to the producers and suppliers. Certainly, looking at it from the point of view of having multiple suppliers enhances our ability to make sure we have secure, reliable, sustainable sources of fuel, and that adds, certainly, to our ability to get our mission done.

Senator THUNE. I, in 2009, sponsored an amendment to the defense authorization bill that allowed for multi-year procurement for alternative fuels. It ended up getting stripped out in conference. But it basically specified that DOD would be able to enter into contracts for up to 10 years if the fuel was cost-effective and environmentally friendly.

I guess right now with the 5-year limitation that you have on contracts, I'm curious in knowing what your thoughts are, and perhaps as well might be able to comment on this—Would it benefit the Air Force and industry if they were allowed to enter into multi-year contracts that exceed 5 years, on the order of 10, 15 and 20 years? And would multi-year procurement authority help to obtain lower energy prices, and increase energy resources?

Mr. YONKERS. Well, the Defense Logistics Agency is the one that buys our fuel for us. But I will comment that I think it will. And we put a legislative initiative in I believe a year or so ago to do exactly what you're talking about, Senator. And I think anytime we can have price stability, it helps in our budgeting, and certainly in our planning and our programming for dollars that we're going to need to operate our Air Force. And it reduces the risk, as I think you've heard from some of our other panel members. And anytime we can do that, I think the cost of fuel is going to drop. But I will certainly defer to the other members.

Senator THUNE. Let me just expand on that if I might, with Mr. Yonkers, Mr. Glover, and Mr. Todaro. As you know, renewable energy and alternative fuel projects are very capital-intensive in terms of investment. Would that sort of expanded multi-year contracting authority for alternative fuels enhance developers' ability to secure financing, to get those types of projects going?

Mr. TODARO. Well, I can certainly speak from our point of view. It depends on the process you use. You know, certainly, longer is better, as cheaper is better. The processes we use tend not to be quite as capital-intensive, so, while 10 years would certainly be welcome, you know, numbers of years longer than that aren't necessary for us.

For us the issue more is a simple indexing system. So, just the same way military and commercial aviation today know what the price of their jet fuel is with high reliability based on the price of crude oil, so we know what the price of renewable jet fuel will be based on the price of our feedstocks. So, if we have long-term contracting authority, that's terrific. But even more important from our point of view is a simple index system. There'll be some years where we may be more expensive than jet, and some years where we may be less. But that non-correlated hedge with domestically grown crops, in my opinion, ought to be something that we're actively pursuing.

Senator THUNE. OK. You, I think, recently were designated one of these BCAP sites. Secretary Vilsack just recently announced four additional projects. And I think you mentioned, in your opening statement, that you were chosen to receive BCAP assistance. BCAP is the Biomass Crop Assistance Program which was authorized in the last Farm Bill.

I'm wondering maybe if you could talk a little bit about how BCAP would be used by AltAir and other companies to specifically develop alternative aviation fuels, how you think it can be used to stimulate growth in the alternative fuels industry, and whether or not you would have any recommendations as we look at the next farm bill, about how to improve that program.

Mr. TODARO. Thank you. That's——

Senator THUNE. There were a lot of questions there.

Mr. TODARO.—right. I was going to say, sort of taking the reverse order.

Senator THUNE. I'm trying to get through this——

Mr. TODARO. Sure. And so, I'll answer them quickly. Yes, we were a recipient. It's a terrific program for companies like ours, because we're out in the farming community. And farmers are conservative people. They make their living based on the weather, and they take risk related to the crops that they have to take. And if someone comes up to them and says, "We'd like you to grow this new energy crop," and the reality is going to be, they'll grow it worse the first year than the second, and worse the second than the third, and by the fourth or fifth year, they'll be experts at it, the same way they are in their regular crops.

That bridge ability, to defer some of their risk, is critical in farmer adoption—whether it's for my program or, really, any others. In terms of, you know, it would certainly be my hope that people understood that, if you're going to create a job an acre, and we're going to do a hundred-million-gallon facility, you're creating a thousand rural jobs in the area of the country where you should most want to create them.

It has a multiplier positive profit-related effect to the Government in ways I've rarely seen in other potential programs under the Government. Thank you.

Senator THUNE. Thank you, Madam Chair.

Senator CANTWELL. Thank you.

Senator LAUTENBERG. Under the—he was here earlier and returned. Very well.

Senator KLOBUCHAR. Of course.

And then we'll go to Senator Klobuchar next.

Senator LAUTENBERG. OK. Thank you very much. I'm sorry to come in for a landing like this second one, I'm——

Senator CANTWELL. We know you're waiting in the wings just for your moment. Senator Klobuchar, would you like to go next?

Senator THUNE. Your side of the isle is so polite.

Senator CANTWELL. Well, if one of you would like to proceed——

[Laughter.]

Senator LAUTENBERG. I promised Senator Klobuchar an ocean. She complained when she first got here that Minnesota didn't have one.

Senator KLOBUCHAR. And I was on the Oceans subcommittee, so that's a little bit of a problem.

Senator LAUTENBERG. I'm sorry to——

Senator KLOBUCHAR. And he said I could come back to the Committee and ask for one. So, there you go.

[Laughter.]

Senator LAUTENBERG. Sorry to take the Committee's time, Madam Chairman.

We're looking at things, Mr. Maurice, that, where we see refusal by the Republicans to pass a clean extension of the FAA authorization. The result—650 people, workers, in New Jersey have been furloughed. And their work on the NextGen system has been, ground to a halt.

Now, what impact do we have with an extended shutdown to get on with our modernization of the air traffic control system? We're looking at runway safety items. And also, to reduce congestion. What are the effects of furlough there now?

Dr. MAURICE. Thank you, Senator, for your question.

Certainly, as you well know, the FAA has 4,000 employees on furlough, and certainly, this is impacting moving forward with the Next Generation Air Transportation System. And certainly, we look forward to a resolution. And being able to move forward with the work is certainly very relevant to the discussions that we're having here today, given that the alternative fuels work is performed under the auspices of NextGen.

And also, keeping in mind that NextGen will make us more efficient—the more efficient you are, the less fuel you need, so the fewer alternatives you need. So, efficiency, and creating an alternative fuel supply work hand-in-hand, and we hope to get back to work as soon as possible.

Senator LAUTENBERG. Thank you.

And I regret that I read exactly what we had, you certainly don't look like a Mr., and we're pleased to have you here.

And I noted before that you come in with a wonderful accent, and the fact that you're here doing what you can suggests that we have to be more inviting in many cases to get the people that we need to do the job, or expand our education process very quickly in this country. Thank you very much.

Now, we're, I think, fair to say, that these rushed arrivals are not very good for orderly process.

Mr. Yonkers, in 2007, 11 former military high-ranking admirals, generals, issued a report that climate change is a threat with the potential to create sustained, natural and humanitarian disasters.

Now, in 2008, General Anthony Zinni warned that if we don't global warming we, he said, will pay the price later in military terms. Can you describe what climate change, how it might affect the Air Force's decisionmaking?

Mr. YONKERS. Well, thank you, Senator. First of all, I read that report from the Center of Naval Analysis many times. I've talked to a number of those general officers. I found the recommendations and observations to be thought-provoking and compelling.

As you probably know, last year, for the first time, climate change energy was put into the Quadrennial Defense Review, so we're now beginning to integrate that into our overall war planning and thought processes.

On a more local level, we're looking at the potential in our master planning areas on our installations of, you know, things like flood control and flood plains.

I would say that in terms of the observations that the general officers and the military advisory board came up with in that report, are, we're just beginning to integrate that into our overall planning.

Senator LAUTENBERG. Ms. Pinkerton, airlines have been creative in finding ways to pass costs along to consumers as, through fees that can account for an additional 20 percent of the ticket price. But with the expiration of the FAA's authority to collect taxes, most of ATA's airline members have elected to pocket the passengers tax revenue instead of lowering ticket prices.

Why aren't these savings being passed on to the consumers?

Ms. PINKERTON. Well, Senator, first I wanted to thank you for your opening remarks, which would, you talked about the importance of the aviation industry in connecting people and goods. And we share your frustration with the situation now.

But that said, I think the message that we want to leave is, we'd like Congress, we're urging Congress to get together to meet and resolve their difference in order to get a long-term FAA bill.

I have to tell you, Senator, the airline industry is sick. It's anemic. In the last 10 years, we've had to shed 150,000 jobs due to our $55 billion worth of losses. Now, we have to be able to cover the cost of flying folks. Our revenues have to be able to cover our costs. And we haven't been doing that in a sustained way.

The reason that's important—it's not just so that we can make a profit, but we have to be able to reinvest—reinvest in creating jobs; reinvest in alternative fuels; reinvest in buying Boeing planes. These things are all very important, and we can't do them unless we make some type of sustained profit over some amount of time, which we simply have not been doing.

I want to leave you with one thing: Yes, some carriers made individual decisions to keep the ticket price, the total ticket price, the same. Customers are paying this week exactly what they paid last week, before the funding lapsed.

Senator LAUTENBERG. Well, I thank you.

And Madam Chairwoman, just indulge for one minute, please.

And that is, that I think a principal thing that can be helpful and, is to let the consumer know very directly what these extra charges are going to be. And I think that we see some airlines doing very well compared to the others. And I think we ought to look to the best and try to take it a little bit easier on the consumers. The fares are going up at a time when incomes have not followed.

So, thank you very much, Madam Chairwoman.

Thanks, Mr.——

Senator CANTWELL. Thank you.

Senator Klobuchar. Senator Klobuchar——

STATEMENT OF HON. AMY KLOBUCHAR, U.S. SENATOR FROM MINNESOTA

Senator KLOBUCHAR. Thank you. Thank you, Madam Chair. Thank you for holding this important hearing.

Air service is incredibly important in my state. We're a major Delta hub. We're also the home of Sun Country Airline. And we also have a number of smaller carriers. I couldn't agree more, Ms. Pinkerton, when you talk about the problems of the airline dealing—right now we've got locations in northern Minnesota where Delta is cutting back service because of jet fuel costs. They are working with us to try to keep, find alternative carriers, and then, just, some of them may simply just be a reduction in one flight. But these are areas that need service. They're International Falls, you may have heard of, which is a thriving town, but it is pretty isolated out there. And we need service to those places.

I think the problems you've identified—we import 61 percent of our oil, making our entire economy vulnerable to swings in the market, and also to politics, oil shortages, upheavals, you name it. And that's why I've, I, in part—one of the most important reasons for developing biofuels. The other is that these are jobs in our country. We can be investing in jobs in the Midwest instead of oil cartels in the Mideast.

And I really appreciate the leadership of the military, Mr. Yonkers, in realizing this, and not only with aviation, but also with some of their other vehicles, their use of biofuels.

I guess my first question would be a local one. We just converted a biofuel plant in southern Minnesota to isobutanol, in Luverne, Minnesota. Is anyone familiar with that? I know that—there are many people in the audience that are familiar with that. But I know it's, one of its uses, in addition to being used for scotch, is also for jet fuel. So, could anyone comment about that? Does anyone—yes, Dr. Maurice.

Dr. MAURICE. Right. I was not aware of the plan in your specific state. But, certainly, the conversion of butanol plants is something that we have been looking at. And in fact, the next class of jet fuels that are being looked at by ASTM International, and that we're looking at—the properties and such—are alcohols-to-jets. And certainly, the butanol conversion to jet would be one of those classes. So, moving that specification through would facilitate and induce the use of the product from your state's new plants. So, that's very exciting.

Senator KLOBUCHAR. And from what I understand, it can also be used, I think, in rubber and other things. So it's very versatile, so that if one market is slow, it can be used in other markets. And that was a big opening that we had down there.

And also, Mr. Yonkers, I'm encouraged by the success, as I mentioned, of the Department of Defense in producing use for alternative aviation fuel from feedstock, or camelina. And I wondered, are you sharing your breakthroughs in results with the wider scientific community? It would seem that creating that domestically produced biofuel source isn't just good for the military directly, but has beneficial, indirect benefits as well.

Mr. YONKERS. Thanks for the questions, Senator. Yes, everything that comes out of our research laboratories is widely dispersed and available to all those in the private sector, whether they're in the biofuels production area, or whether they're in the aviation and commercial aviation sector.

And I would just add one thing in the remarks to the alcohol-to-jet. I mentioned that this is probably, from my perspective, one of the more promising pathways, because you can use waste product. Anything that has cellulose in it is available for this particular pathway. So——

Senator KLOBUCHAR. I would think you would think that continuing that cellulosic tax credit will be important going forward.

Mr. YONKERS. That will be a decision for you all to make.

Senator KLOBUCHAR. All right. Very good.

Mr. Todaro, you mentioned that you predict that you'll be able to produce the camelina—am I saying that right?—based on aviation fuel at or slightly below the cost of petroleum-based jet fuel. And is there a possibility, with the expansion of the aviation biofuel industry, that the cost could fall, making it even more profitable and easier to produce?

Mr. TODARO. Well, let me be clear. That at the moment is somewhat aspirational. I think that Senators on the panel have disclosed some of the prices for the early test batches that we've done. There is something called Zeno's paradox. We have been able to have our prices every time we've made a delivery to the Department of Defense or commercial aviation. That trend will continue for quite some time.

To know whether or not it costs more or less than jet fuel, you need to be able to answer questions that none of us can answer—What's the price of crude oil? If you can tell me that, I can tell you whether or not we're likely to be able to beat it.

But I can tell you that these, the fuels we're working on today, at any reasonable scale, will be much closer to jet prices than you would think. So, you're not talking about something that's twice as expensive, or even 50 percent more. You're talking about something that, at scale, should be competitive.

Now, look, we're a business. If we produced it at prices cheaper than jet, it will always be priced at jet, because that's what sets the market. But in conversations we've had with airlines, the most progressive among the airlines are looking at negotiating with us so that they can actually lock in some of the potential savings on fuel that we think we'll be able to deliver in the next 5 years. And those carriers will receive that economic benefit.

Senator KLOBUCHAR. Thank you.

And then, last, Ms. Pinkerton—maybe Mr. Glover can answer this too—but, the role of speculation that you see. I know I've talked to Richard Anderson, the CEO of Delta many times about this. And he's very concerned about that and how that affects their projections, and affects their ability to fly people. Could you talk about speculation? Because we've been trying to get the Commodity Future Trading Commission to put forward the rules.

Ms. PINKERTON. That's correct. And I complimented Senator Cantwell earlier for her leadership, and we appreciate your leadership as well on this issue.

As you know, there's 17 times more speculative activity in the oil trading market than there is bona fide physical hedgers like us. We supported the provisions in the Dodd-Frank Act, that essentially gave the CFTC and mandated the CFTC to apply position limits, to try to reduce excessive speculation in the market. That's taking a frustratingly long time for those rules to come out.

We continue to work together with many other impacted industries who are, unfortunately, like consumers, on the losing end of excessive speculation.

Senator KLOBUCHAR. Mr. Glover, do you want to add anything? You don't?

Mr. GLOVER. Nothing to add.

Senator KLOBUCHAR. OK. Thank you.

Senator CANTWELL. Thank you, Senator Klobuchar.

And I thank the panel very much for their testimony, and particularly because you came with specific recommendations. We always appreciate that. And paths to follow, particularly, Mr. Glover, Mr. Todaro, thank you so much for your leadership on this issue. I appreciate it very much, as representative of Washington State. And thank you to this panel.

I'm going to call up the next panel, because we certainly want to hear from them as well.

And I just want to point out that we will leave the record open for further questions, if you'd be so happy to, anything that members submit to you, be happy to have your answers supplied to those questions. Thank you all very much.

We're next going to hear from Mr. John Plaza, President and Chief Executive Officer of Imperium Renewables; Mr. Richard Altman, Executive Director of Commercial Aviation and Alternative Fuels Initiative; and Ms. Judy Canales, Administrator for Rural Business and Cooperative Programs with the U.S. Department of Agriculture.

So, if the second panel could make their way up to the dias there.

And if we could make a somewhat orderly transition out of the room, it would be much appreciated.

All right. Ms. Canales, we'll start with you. Thank you very much for being here.

STATEMENT OF JUDITH CANALES, ADMINISTRATOR, RURAL BUSINESS SERVICE, UNITED STATES DEPARTMENT OF AGRICULTURE

Ms. CANALES. Chairwoman Cantwell, Ranking Member Thune, and members of the Subcommittee, I appreciate the opportunity to appear before you today and testify on the USDA's role in addressing the needs, challenges, and alternatives to aviation fuels.

I have submitted full testimony for the record, and will briefly summarize my comments before you today.

In July 2010, the U.S. Department of Agriculture, Air Transport Association of America, and the Boeing Company signed a resolution formalizing our commitment to work together on a "Farm to Fly" effort to accelerate the availability of a commercially viable, sustainable aviation biofuel industry in the United States, increase domestic energy security, establish regional supply chains, and support rural development.

This effort was created in response to President Obama's directive to meet our obligations under the Energy Independence and Security Act of 2007 to produce 36 billion gallons of biofuel annually by 2022.

Since July 2010, USDA has joined with the U.S. aviation community and the United States Departments of Energy, Transportation, Defense, and Commerce to examine how aviation biofuel can become, in the near future, an economical and environmentally preferred alternative to petroleum-based jet fuel.

Our coalition's strong commitment of resources to research, development, demonstration and deployment, using public sector leadership and financial incentives will bring production online. This commitment includes the creation and implementation of programs and incentives to assist American farmers and foresters in the selection and cultivation of energy crops and feedstocks that can be converted into affordable and sustainable aviation biofuels.

USDA has several programs authorized in the 2008 Farm Bill which support the Farm to Fly effort across the supply chain. For my oral testimony here today, I will focus on two of these complementary programs.

The Biomass Crop Assistance Program, known as BCAP, administered by my sister agency, the Farm Service Agency. BCAP is the only energy program primarily dedicated to the expansion of the diversity of feedstock for commercial conversion. The program has demonstrated through project area proposal submission and matching payment distribution that demand for feedstock support exists. Just this past Tuesday, as you well know, the U.S. Department of Agriculture announced a BCAP project area for one of the companies that was in the earlier panel today, AltAir, a company formed to manufacture bio-based jet fuels from camelina grown on up to 50,000 acres in Oregon, Washington and Montana.

A program that I operate in my agency, the Biorefinery Assistance Program, Section 9003 of the 2008 Farm Bill, is administered within the Rural Business Cooperative Programs. This program provides loan guarantees for the development, construction and retrofitting of commercial scale biorefineries that produce advanced biofuels. To date a total of $390 million has been awarded in loan guarantees, including one to Sapphire Energy for $54.5 million.

The Sapphire Energy facility will produce algal oil to be refined into aviation biofuel.

Additionally, we are currently reviewing at this time 10 applications in the 9003 program which will include some projects that pertain to aviation biofuel production. The total of 10 projects requests roughly $1 billion in loan guarantees. They are competing for $463 million that are available for this program today.

But why commercial aviation? Among transportation modes, aviation is unique in its complete dependency upon liquid fuels. This biofuel production will support our Nation's rural economy, create employment opportunities, and provide stability to the aviation industry with its 19 billion gallon per year commercial jet fuel market. There are currently dozens of U.S. aviation biofuel projects waiting in the wings in various stages of development.

As with many emerging technologies, a strong governmental role is essential to build the financial bridges needed to assist the aviation biofuel industry through its nascent development. Expediting the commercial production of aviation biofuel sustains not only the aviation industry, but also strengthens the competitiveness of the American farmer and creates thousands of jobs across America by processing the biomass and building the logistical infrastructure that is needed to support renewable fuel processing economies.

Thank you for your time and your leadership, Madam Chair and members of the Subcommittee.

USDA is committed to promoting aviation fuel through the Farm to Fly effort, and will continue to promote rural development and job creation to support this growing effort.

Thank you.

[The prepared statement of Ms. Canales follows:]

PREPARED STATEMENT FOR JUDITH CANALES, ADMINISTRATOR, RURAL BUSINESS SERVICE, UNITED STATES DEPARTMENT OF AGRICULTURE

Chairman Cantwell, Ranking Member Thune and members of the Committee, I appreciate the opportunity to appear before you today and testify on the USDA's role in addressing the needs, challenges and alternatives to aviation fuels. In July 2010, the U.S. Department of Agriculture, Air Transport Association of America and The Boeing Company signed a resolution formalizing their commitment to work together on a "Farm to Fly" initiative "to accelerate the availability of a commercially viable sustainable aviation biofuel industry in the United States, increase domestic energy security, establish regional supply chains and support rural development." In addition, USDA has an MOU with the FAA on the development of research related to aviation biofuels, but I will defer to the FAA to cover the details of that arrangement.

The Opportunity

In his State of the Union address on Jan. 25, 2011, President Obama reaffirmed the administration's commitment to government investment in clean-energy technology research, development, and deployment, "an investment that will strengthen our security, protect our planet, and create countless new jobs for our people." The President's remarks underscored the "promise of renewable energy," building on his pledge to develop a commercially viable biofuels industry in America.

Just a year earlier, on February 3, 2010, President Obama had announced a series of steps that the administration was taking to boost biofuels production in the United States. The Biofuels Interagency Working Group released a report spelling out ways to promote the development of the biofuels industry in the United States in connection with the Energy Independence and Security Act target of 36 billion gallons per year of U.S. biofuels production by 2022. The report, "Growing America's Fuel," laid out the situation and called for "an outcome-driven re-engineered system." The strategies include supporting the development of first- and second-generation biofuels and accelerating the development of third-generation biofuels—including aviation fuels. These strategies were further highlighted in the USDA Biofuels Strategic Production Report and regional roadmap released in June of 2010.

Farm to Fly

As a result of the Farm to Fly initiative, the U.S. aviation community has come together with government stakeholders, including USDA, the Department of Energy, the Department of Transportation, the Department of Defense, and the Department of Commerce to express unified support for the President's goals of environmental stewardship and energy security. The coalition was formed to help aviation biofuels become an economical and environmentally preferred alternative to petroleum-based jet fuels. In pursuit of that goal, the initiative is coordinating exchanges of information to inform research and development activities while capitalizing on the existing efforts across the government to spur innovation. For example, DOE's research, development, and deployment activities on technologies for biomass handling and conversion to fuel, power, and products are complementary to USDA's activities and are increasingly focused on hydrocarbon advanced biofuels such as jet fuel.

Why Commercial Aviation?

In 2010 the U.S. produced approximately 13 billion gallons per year of biofuels, mostly corn grain ethanol. However, in comparison to the investments made in surface transportation fuels like ethanol and biodiesel, government efforts to date have not emphasized research, development, or commercialization of alternative fuels for aviation. Yet, commercial aviation is a central contributor to the modern American economy and, among transportation modes, aviation is unique in its complete dependence on liquid fuels.

Moreover, rather than delivering this fuel to tens of thousands of gas stations and convenience stores around the country, the largest 35 U.S. airports account for about 80 percent of the jet fuel used by commercial aviation. Thus, if aviation biofuel producers can deliver to these 35 airport "gas stations," they have access to virtually the entire 17-to-19 billion gallon-per-year commercial jet-fuel market. The production of environmentally preferred aviation biofuels by U.S. companies also will support the President's goal of reducing imported oil by 1/3 by 2020 and increasing U.S. exports to support our Nation's rural economy and to win the future.

Synergy with the U.S. Military

The U.S. Air Force is working to have one-half of its jet fuel be nonpetroleum-based by the year 2016. The Department of the Navy (DON) has announced a goal of supplying 50 percent of its total energy consumption from alternative sources by 2020. USDA and the DON announced on January 21, 2010 that Secretary Vilsack and Secretary Mabus signed a Memorandum of Understanding (MOU) to help meet these goals and encourage the development of advanced biofuels and other renewable energy systems. The military services have implemented robust programs to reach these goals. Significant collaboration and coordination on research and development and on fuel approval and deployment by commercial aviation and military efforts has allowed for significant mutual benefit and more rapid progress.

USDA Programs which can Support Aviation Biofuel

There are several programs related to alternative fuels production under the 2008 Farm Bill. These programs provide additional opportunities for access to credit in rural America and jump-start the biofuels industry. The following *2008 Farm Bill* programs were designed to support the biofuels industry and have contributed to our efforts.

- *Biorefinery Assistance Program (Section 9003 of the 2008 Farm Bill).* The Biorefinery Assistance Program (BAP), administered by USDA-Rural Development, provides loan guarantees for the construction or retrofitting of rural biorefineries to "assist in the development of new and emerging technologies for the development of advanced biofuels . . . made from renewable biomass, other than ethanol from corn kernel starch." It does so by guaranteeing up to 90 percent of a private loan (not to exceed $250 million) to construct first of kind/scaled to commercial level or retrofit commercial-scale biorefineries producing advanced biofuels. To date, a total of $390.1 million has been obligated in loan guarantee authorities to leverage an estimated $1.5 billion in total project costs toward the construction of commercial scale advanced biofuel facilities, including Sapphire Energy for $54.5 million which will have the capability of producing aviation biofuel from algal oil. The Rural Business-Cooperative Service Agency is currently reviewing 10 applications, including multiple aviation biofuel projects, for the remaining $463 million available at the program level.

- *The Bioenergy Program for Advanced Biofuels ("BPAB"—Section 9005 of the 2008 Farm Bill)*: BPAB gives the Agriculture Secretary broad discretion—and $300 million—to create a program to provide production payments to eligible advanced biofuel producers, "to support and ensure an expanding production of advanced biofuels."
- *Biomass Crop Assistance Program ("BCAP"—Section 9011 of the 2008 Farm Bill)*: BCAP is the only energy program primarily dedicated to the expansion of the diversity of cellulosic feedstock for commercial conversion. The program has demonstrated, through project area proposal submission and matching payment distribution, that demand for feedstock support exists.

  Just this last Tuesday, USDA announced a BCAP project area for AltAir, a company formed to manufacture bio-based jet fuels from camelina grown on up to 50,000 acres in Oregon, Washington and California.
- *Crop Insurance Coverage for Energy Crops (Section 12023 of the 2008 Farm Bill)*: The *2008 Farm Bill* directed the Risk Management Agency (RMA) to research and develop "a policy to insure dedicated energy crops," defined as crops "grown expressly for the purpose of producing a feedstock for renewable biofuel, renewable electricity or biobased product, and is not typically used for food, feed or fiber." RMA has recently awarded a contract to conduct this research.
- *National Institute of Food and Agriculture (NIFA)*: NIFA was created by the *2008 Farm Bill* to fund competitive, peer-reviewed research efforts. For example, the Biomass Research and Development Initiative (Section 9008 of the *2008 Farm Bill*) made $118 million available for uses that include advanced research on feedstock development, biofuels, and bio-based product development and biofuels development analysis. In addition to BRDI, NIFA offers a series of Sustainable Bioenergy grants through its Agriculture and Food Research Initiative, and also operates the Plant Feedstock Genomics for bioenergy program, both competitive grant programs support research and development of bioenergy.

Flying into the Future

Expediting the commercial production of aviation biofuels will strengthen those elements of the agricultural sector involved in the growth of biomass, the "green" technologies that process the biomass, and those who build the logistical infrastructure that is needed in select areas. Over time, the investments made today will lessen our reliance on petroleum-derived fuels. The Farm to Fly effort aims to "accelerate the availability of a commercially viable, sustainable aviation biofuel industry in the United States, increase domestic energy security, establish regional supply chains and support rural development." We look forward to working with the Congress to reach these goals and to fly into the future.

Senator CANTWELL. Thank you very much.

Next, Mr. Altman. Welcome. Thank you for your testimony.

STATEMENT OF RICHARD L. ALTMAN, EXECUTIVE DIRECTOR, COMMERCIAL AVIATION ALTERNATIVE FUELS INITIATIVE (CAAFI)

Mr. ALTMAN. Thank you, Madam Chair. Thank you for the invitation to the Commercial Aviation Alternative Fuels Initiative.

Those testifying before you earlier are all parties to the CAAFI coalition, a supply chain initiative. Boeing, FAA and ATA are key sponsors of the Commercial Aviation Alternative Fuels Initiative. AltAir and Imperium are prominent among the 60 fuel company stakeholders and are at the cutting edge of that group in advancing deployment. Together in, some dozen U.S. biofuel companies are responsible for nearly 80 percent of the aviation biofuel programs now resident in a dozen countries around the world.

The Research Lab at Wright Patterson Air Force Base in Ohio predates CAAFI and remains at the core of aviation research and qualification efforts. CAAFI would never have started and will not be completed in a timely manner without Air Force support.

I would like to single out two individual and groups that are part of our activity. First is to cite again the FAA Office in Environment and Energy, Dr. Maurice, and very specifically, leaders of our process, Nate Brown at FAA and Dr. Kristen Lewis at DOT and Volpe. If my right arm seems dangling off here at the moment, it's because Nate isn't here because he was furloughed as part of the current funding lapse. So, our number one goal is to get the FAA and the 4,000 employees—at least, for me, the dozen in Dr. Maurice's office—back to work.

The centroid of sustainable aviation fuels with commercially viable off-take agreements between buyers and sellers. I'm pleased that in the audience today are representatives of not only ATA, but also American Airlines and United Airlines, who are sitting behind me here. Together, U.S. aviation comprises over 85 percent of total U.S. jet fuel demand.

Critical to the progression of alternative fuels is our fourth sponsor—the airports through which these fuels do flow. The emerging industry offers unique opportunities and projects occur now in your region. In Spokane, Sea-Tac, Portland, and also other locations such as in New York City. The Aerotropolis of Detroit is another area where we have fuel projects. These are just a few examples.

The CAAFI coalition are active in some 20 states in the United States. The progress is reflected in the President's energy policy announcement of March 30 targeting our industry. Aviation peers have also recognized CAAFI in that the progress that has been made in aviation alternative fuels. CAFFI was awarded the Air Transport World's Industry Service award in 2010.

As for current projects, it's not a question, as Mr. Glover pointed out, of if, but it's now when and how we progress. So, that is my stated focus, and as the Chair has asked, we have five specific recommendations in this regard. One, increase supply; two, drive down costs; three, reduce environmental uncertainty; and four, ensure commercial success for the suppliers to the aviation enterprise.

First, and most critical, and within the authority of this committee, is that we now leverage the new ASTM protocols to enable qualification of fuels emerging from catalytic processes—you've heard about alcohol-to-jets—synthetic biology, and from pyrolysis as well as other technologies that are emerging at a fast pace. Lipid seed crops, such as what Mr. Todaro discussed, while a great start, simply cannot assure adequate supply in the target timeframes. Such success requires the support of FAA, NASA, the Air Force and the Navy.

The Committee's funding of advanced biofuel programs at FAA last year is a good start, but it is only a start. As one who's spent 39 years in this career, I am intimately familiar with the challenges of reducing cost—I was in the manufacturing sector. No new technology—not computers, not aircraft—had mature costs at the very initial product that was launched. Success comes from learning, and the needed infusion of production technology to enable acceleration of that learning through accepted methodologies.

Such programs to reduce are in place in USDA, in coordination with FAA, and at DOE. We are helping, and we have seen these programs as quite effective.

The third are questions of financing. Clearly, USDA's suite of products under Section 9003, loan guarantees and BCAP, are the kind of near-term and finite lived programs that need to continue to ensure jobs and energy independence. And, thank you, Administrator Canales, for those programs.

Beyond first-of-a-kind funding in USDA we simply need to emphasize allowing institutions that lend overseas as part of their charter to support aviation biofuel projects. Aviation biofuels are a real industry that should be supported by our overseas financial support from institutions that support U.S. companies, and, by "Invest in U.S." programs just started at the Department of Commerce.

Fourth, certainty and quantification of certifying environmental performance is extremely critical for both carbon life cycle and sustainability. In this regard, aviation has benefited from a comprehensive peer review analysis of carbon life cycle executed by the Air Force, DOE, and the FAA. We need to continue that process of adding to the data base, as processes and feedstocks mature, and to ensure full coverage.

Last, the tools that we use to assess alternative fuel projects and permit their analysis by airlines and airports, as well as by fuel producers and stakeholders from the agricultural community, as defined by ACRP, need to be kept up to date.

So, in closing, I would again like to thank you, Madam Chair, and the Subcommittee members for giving CAAFI coalition as a whole the opportunity to present its views today.

[The prepared statement of Mr. Altman follows:]

PREPARED STATEMENT OF RICHARD L. ALTMAN, EXECUTIVE DIRECTOR, COMMERCIAL AVIATION ALTERNATIVE FUELS INITIATIVE (CAAFI)

Madam Chair, Senator Thune, and members of the Subcommittee:

Thank you for the invitation to the Commercial Aviation Alternative Fuels Initiative (CAAFI) and me as its Executive Director to testify at today's hearing on "Aviation Fuels: Needs, Challenges, and Alternatives."

CAAFI is a U.S. based supply chain coalition that has been a leader in the push toward the development and deployment of Aviation Alternative Fuels having Economic, Environmental and Security of Supply benefits.

Formed on the basis of a Boeing hosted meeting of the Transportation Research Board held in Seattle in May 2006 CAAFI was founded to propel commercial aviation to a position of leadership in the quest to secure, clean, and economic alternative transportation fuel supplies. In forming CAAFI its founders addressed the challenge of being a- a minority player (10 percent) of the transport market by fuel use. CAAFI sought to transform aviation from an afterthought in alternative fuel thinking to a "first mover". As a unique minority we needed to do so or run the risk of being left without alternatives. We built CAAFI using aviations inherent attributes (concentrated distribution, unified, rationale, technically skilled customers) and turning our unique characteristics into strengths (*e.g.,* dependency on liquids, high qualification barriers).

Those who testified before you earlier are all parties to the CAAFI supply chain coalition. . . . Boeing as a leader among Aerospace Industry Association manufacturers, the FAA office of Environment and Energy, and the Air Transport Association of North America are key CAAFI sponsors and founders.

Altair along is prominent among some 60 fuel company stakeholders and is at the cutting edge of that group in advancing deployment. Together some dozen U.S. biofuel suppliers have been responsible for nearly 80 percent of the aviation biofuel programs now resident in a dozen countries around the world.

USAF through its Research Lab at Wright Paterson AFB in Ohio predates CAAFI and remains at the core of aviation research and qualification efforts. None of what you are hearing today would have started without the Air Force's outstanding contributions at a time when there were few believers in aviation. The job of CAAFI would never have started and will not be completed in a timely manner without the Air Force.

CAAFI sponsor, AIA and its members Boeing and engine manufacturers General Electric, Pratt and Whitney and Honeywell, in combination with the FAA have led CAAFI Certification team in a way that has transformed an ASTM approval process. The process that took a decade and ten's of million dollars to qualify fuel from a single producer and process location has now evolved into a robust methodology that has produced qualification of two new process categories in the 2009 to 2011 period for global supply. In so doing, time to qualification has been cut by two thirds. Qualification costs by similar amounts. Together we have enabled the formation of an Aviation alternative fuels supply industry.

The FAA Office of Environment and Energy in addition to the discrete accomplishments outlined by Dr. Maurice has been the focal point for some seventeen different government agencies having complementary responsibilities from crop growth to technology formation and through Commercial program development. In my 44 years of industry participation and government interaction the alternative aviation fuels initiative within government has had both the most dimensions and been the most successful of any intra-governmental activity that I personally have been witness to. Particular credit goes to FAA/DOT members of CAAFI leadership team, Nathan Brown of FAA and Dr. Kristin Lewis of DOT/Volpe. Nate and Kristin have been instrumental in forming this cooperation and the programs which they include, and, deserve much of the credit for that success.

The Air Transport Association of North America represents 90 percent of North American Airlines and nearly a quarter of aviation fuel consumption worldwide. Their roles at the top of the supply chain as buyers (ATA) has now progressed to the forefront as the issue of sustainable alternative fuels moves to the deployment phase. Clearly the centroid of the sustainable aviation alternative fuels resides with commercially viable off-take agreements between buyers and sellers at this point in time. I am pleased that in the audience today are representatives of American Airlines, United Airlines as well as ATA. Separately AA and UA purchase as much fuel as the U.S. Air Force. Together U.S. commercial aviation comprises over 85 percent of total U.S. Jet fuel demand. This year global commercial aviation will consume over 70 Billion gallons of jet fuel.

CAAFI member Airlines are committed to motivate development and deployment of alternatives that are both environmentally advantaged and that provide for improved economics. ATA and its members have been active in cultivating agreements among airline buyers for fuel purchases that include four MOU's that form the basis for long term offtake agreements on fuels ranging from Altair's camelina based fuel to a recent agreement with MSW to Liquid suppler Solena in Gilroy, California. Along with Boeing, ATA is a CAAFI sponsor signatory to the July 2010 "Farm to Fly" agreement that is accelerating and focusing Biofuels initiatives for Commercial Aviation. Across the public/private purchasing communality ATA are teamed with DLA Energy to afford producers the prospects for alignment among military and Commercial purchasers.

Together with CAAFI Stakeholders in some 20 U.S. States ATA are helping to organize State initiatives which are linking agriculture, energy, business development and aviation interests in these states. For these states ATA member airlines provide buyer focals to help focus state interests around real and substantive buyers. ATA and its global partner IATA are committed to seeking Carbon neutral growth for the airlines globally by 2020.

Also critical to the progression of alternative fuels are the Airports which these fuels flow through. The emerging industry can be seen to offer unique business opportunities deployment of these fuels can offer airport business growth and environmental gain at the same time, and also can afford similar benefits for the communities in which they reside. 80 percent of all jet fuel flows through some 35 airports in the U.S. Such concentrated distribution offers a unique opportunity for Airport business interests and for fuel suppliers alike.

ACI North America airports are becoming increasingly important and significant participants in the progression toward aviation alternative fuels. Seattle, Portland, Spokane, in SAFN have been joined by ACI members large and small in recent months. The Port of New York Authority has been working with CAAFI stakeholders to evaluate MSW to liquid opportunities for its supply. Detroit through DTW and three universities are evaluating the opportunity to turn the Aerotropolis linking the airport with other state owned lands into the origin of home grown biofuels opportunities both in the Aerotropolis and subsequently into bordering communities. Such developments offer economic and environmental benefits as part of the revival of that region.. Smaller airports such as Tulsa are moving to explore new options with local entrepreneurs and universities.

As the evaluation of specific fuel projects requires extensive and quantitative analysis the Airport Cooperative Research Program . . . under the Transportation Research Board jurisdiction, with FAA and ATA support, has launched a series of three programs to lay out methodologies to calculate benefit and cost assessments of individual projects. These projects offer detailed analyses of emissions reductions for small particles, aviation user and airport planning tools, and a just launched multi-modal assessment tools that will allow the 50 percent or biofuel plant output that is not jet fuel to be evaluated for distribution potential through the airport and its clients. Many additional airports beyond those mentioned serve on ACRP the panels overseeing these projects. New candidates are coming forward all the time as momentum grows toward sustainable, secure aviation alternative fuel deployment.

The potential for a whole new business for airports, capitalizing on the concentrated distribution afforded by our industry and producing not only economic but environmental gains could well be a major "fringe benefit" of the emerging aviation alternative fuels infrastructure development. Biofuels programs supported by USDA, DOE and State and Local communities are an addition to those programs supported by the Aviation sector and offer a once in a generation opening for our industry to lead the country toward a new and promising future.

Collectively Aviation Alternative Fuels have shown great progress as has been reported to you today.. This progress is reflected in the President's Energy Policy announcement of March 30 targeting our industry. That decision is backed by a poll by *Biofuels Digest* who in October, 2010 polled 40 percent of the Biofuels industry indicated that there expectation was that there would be 1 Billion gallons of biojet produced annually by 2020. Business publications such as "the *Economist*," in that same time period, suggested that the future of biofuels was now metaphorically, "looking up." Aviation peers have also recognized CAAFI and the progress that has been made in Aviation Alternative fuels through the award of Air Transport World's prestigious Joseph S. Murphy Industry Service award in 2010 to CAAFI and its sponsors, FAA, ATA, ACI–NA and AIA.

Indeed much has been done. Industry/government in partnership are clearly focused, unified and dedicated to making an environmentally friendly, economic and secure future for aviation alternative fuels industry practical near term. Such success enables us to more narrow the focus of needs, and challenges for alternatives. This charge from the Committee for this session is the clear focus for the remainder of my remarks.

Aviation accomplishments with favorable environmental, economic and security of supply implications places us passed a key inflection point. The issue is no longer what or if aviation alternative alternatives can serve as a spinoff of an energy supply sector such as oil. The needs and challenges are now when and how aviation leads not only aviation but the transport biofuels sector to success are the focus of the needs and challenges that I personally would like to focus upon as areas of emphasis across all CAAFI sponsor and stakeholder interest.

In this regard I would like to focus on five areas that will increase supply, drive down cost, reduce environmental uncertainty, and ensure commercial success for the suppliers and aviation enterprise. With success in all five of these areas in parallel aviation will not only lead the transport sector but the Nation to create a tremendous economic and environmental asset. That is a role we have played many times.

First, it is critical and within the authority of this committee that we leverage new ASTM protocols to enable qualification of fuels emerging from catalytic processes, synthetic biology and from pyrolysis as well as other technologies that are emerging at a fast pace. Lipid seed crop production (HRJ, HEFA) while a great start simply cannot provide an adequate supply in the target time frames. Such success requires ongoing support of FAA, NASA, USAF, Navy. With the breadth of the opportunity so large cooperation with international partners with proven capability and with whom both the agencies and private sector have considerable experience working is in order. Such an effort can be guided via an upgrading of our R&D roadmaps and use of our globally accepted risk management (Fuel Readiness level) methodology to ensure aligned and complimentary efforts with little overlap. With the possibility of three paths at a minimum to be pursued in parallel through the 2013–15 we can vastly increase candidate supplies to include cellulosic sources that grow in lands that do not conflict with food production much as the targeted seed crops do. Commitments that offer one year of such research while a good start, without needed follow-up from the inevitable questions that they produce will not provide the needed outcome in my personal view. We simply need an ongoing commitment to Advance fuels R&D through current funding sources. This committee's funding of the Advance Biofuels program at FAA last year is a good start.

Second, is the question of cost. As one who spent the first 39 years of his career I am intimately familiar with challenges of production learning. No new technology, not computers, not the material we use today in our households and very especially aircraft with which this room is most familiar with ever produced learned out, mature, costs at the outset of production. Those citing the cost of limited supplies in the short term as the true state of affairs should ask computer makers what the first computer cost, or perhaps aircraft makers what the fist of a kind aircraft cost. Success comes from learning and the needed infusion of production technology to enable acceleration of that learning through accepted methodologies.

No one is better at that than the agriculture sector. Working with the CAAFI R&D community and FAA we have now created a Feedstock Readiness level methodology via a unique intergovernmental MOU with USDA. Aviation systems technology and learning combined with well structured USDA efforts at its new research centers combines the best know how in the world to speed learning for fuels from oilseed sources. Qualified HRJ/HEFA fuels are dominated 80 percent or more by feedstock costs.

For other advanced processes is driven through process cost reduction. This is the sanctuary of DOE biofuels research. I personally have seen in detail DOE's plan to attain $2 a gallon cost for pyrolysis oil. It depends primarily on increasing the life of catalysts used in pyrolysis production to that of similar catalysts in oil refineries. The DOE programs have finite plans with check points along the way. It is a challenge to attain similar life under higher pressures and temperatures but it is a challenge that I know aviation engine manufacturers successfully addressed during my tenure in manufacturing sector. It is doable and the plans are sound ones.

Third are questions of financing. Clearly USDA's suite of products under section 9003 loan guarantees and the Biomass crop assistance programs (BCAP) are the kind of near term and finite lived programs needed to ensure jobs and energy independence for America. Having been a part of programs in Europe and those that are seeking to be formed in other places I can advise that they are truly the envy of the world. Such incentive-based approaches clearly produce positive results. Without these first of a kind programs private sector participation, particularly after the crisis that we just encountered in the financial sector will not keep pace with technical achievements. Retaining such programs will enable reduction in the costs of protecting access to foreign supplies. That cost is a much smaller order of magnitude in dollars and human sacrifice and as such is an excellent national investment.

Beyond first of a kind funding via the USDA programs a simple emphasis in allowing institutions that lend overseas as part of their charter to encourage investments from overseas in our country to fund aviation biofuels projects as real commercial efforts are in order and are now enabled for USDA programs. Aviation biofuels is a real industry that should be supported by our overseas financial support institutions that support U.S. companies, and by "Invest in U.S." programs just started by the Commerce Department. CAAFI fuel stakeholders were invited by DOC Undersecretary Sanchez and Assistant Secretary Nicole Lamb-Hale to engage in discussion of these matters and other follow-on measures to what was a highly successful Paris Airshow CAAFI showcase. We are proceeding to develop that direction with DOC and our government stakeholders.

Fourth, certainty and quantitative means of certifying environmental performance of future facilities for both carbon life cycle and sustainability is a critical need for fuel producers and buyers alike. In this regard Aviation has benefited from a comprehensive, peer reviewed analysis of carbon life cycle outcomes executed by the Air Force and DOE with the aid of quantitative carbon life cycle analysis from ground to wake funded by FAA Partner. We need to continue the process of adding to the database as processes and feedstocks mature and to ensure full coverage. In addition the benefits of carbon savings need to be attributed equitably to all who purchase the fuel in calculating economic benefit.

Lastly, the tools that we use to assess alternative fuels project and permit their analysis by airlines, airports as well as fuel producer and grower stakeholders that are being defined by ACRP need to be kept up to date. While the ACRP projects, of which I personally am a party to constructing, are excellent in scope and purpose the methodology must be fed by an up to date database of information regarding feedstocks and processes. Currently, ACRP executes its work at a given point in time. Shelf life of the data bases is inversely proportional to the speed that the space is growing. The aviation alternative fuels space is growing rapidly. A mechanism should be found to keep these tools up to date and relevant for the intended use.

In closing, I would again like to thank you, Madam Chair, and the Subcommittee members for giving the CAAFI coalition as a whole the opportunity to present its views today. I would also like to take the opportunity to invite the members of the Committee and your staff to gain more comprehensive exposure to the both our Coalition and our sponsors and stakeholders through attendance at the CAAFI bi-annual meeting and CAAFI Expo to be held at Georgetown on Nov. 30 and Dec. 1. At Georgetown, our full complement of sponsors and stakeholders will demonstrate the breadth and depth of what has been achieved and what, with the assistance of the Committee, we can achieve in the future.

Senator CANTWELL. Thank you, Mr. Altman. And thank you for your work on this association. It is a broad coalition, and we appreciate that.

Mr. Plaza, welcome. Thank you for being here, also from Washington State. We very much appreciate your testimony.

STATEMENT OF JOHN PLAZA, PRESIDENT AND CEO, IMPERIUM RENEWABLES, INC.

Mr. PLAZA. Thank you, Madam Chair.

My name is John Plaza. I'm the President and CEO of Imperium Renewables. We're headquartered in Washington State.

I greatly appreciate the chance to appear before the Subcommittee today to discuss commercialization of renewable jet fuels, as well as the importance of long-term purchase commitments by the Department of Defense.

I also want to thank you, Senator Cantwell, for your continued support for the biodiesel industry. As you know, that's critical to our success.

A little bit about Imperium Renewables, who we are. This is our facility here. We own and operate one of the largest biodiesel facilities in the world—it's 100 million gallons of capacity. We produce using exclusively canola from the northern tier of the United States, as well as Canada. We rail and ship it in, and have the ability to ship biodiesel around the world.

We spent about $90 million building this facility. It was a green field state-of-the-art facility, our own technology. We currently employ about 42 people in the company, five of which, who are veterans who've returned from Iraq and Afghanistan and found a job in our company, and they're very excited about that for a multitude of reasons.

Since this is a Commerce committee, I wanted to point out a few economics. One of the things that we hear a lot are the value benefit of renewable energy in our Nation. Our company alone has invested $135 of direct economic benefit into the Pacific Northwest, namely, Washington State. We have done that through both labor, payroll, taxation, investment to vendors, construction, and a number of different things.

Before I really get into what we think we need to do, I want to tell you a little bit about the background of the company. We've been around since 2005. Prior to starting the company I was an airline pilot. I flew for Northwest Airlines, a number of other airlines. I've flown everything from small bush planes in Alaska to 747s across the Pacific. I have a deep understanding of the critical importance of fuel quality, the safety of aviation, and the importance of the price of fuel for aviation.

We're the first commercial producer of renewable jet fuel. We partnered with Virgin Atlantic, General Electric, and, most importantly, Boeing, in a demonstration on a 747 from Amsterdam to London in 2008.

What we want to do—we have a state-of-the-art facility. As you see in this picture, there's a number of acreage, or, there's a large area of acreage next to us that we can build additional capacity. We'd like to build a renewable jet fuel facility adjacent to our existing site. We think renewable jet fuel using vegetable oils and the HRJ process is ready for commercialization now.

Obviously, you've heard that ASTM approvals happened already this year for up to 50 percent blends. This means it's allowed in any aircraft globally. This really takes the industry out of the R&D phase that it has been in for the last 4 years and into the commercialization phase.

By building this facility at our existing site, we'd create over 300 construction jobs over a 3-year period, and increase our workforce by an additional 50 permanent employees, and since we're in a rural area, jobs are hard to find, and a family wage job is a big deal in our community.

With the construction and operation of this new facility, an additional $250 million will be invested into the state with this project alone. Once in operation, there would be another 2,000—excuse me—$20 million of annual direct economic benefit to the region as well.

We have an aggressive plan. We think we can produce significant amounts of jet fuel by 2014 using vegetable oils that are commercially available now. We will certainly continue to provide additional jobs. We'll work toward the effort to displace over $1 billion that we, as a Nation, borrow from China every day to import petroleum from overseas. This is a huge drain of our economy, and we think this is an industry that can help reduce that. And it certainly increases our military's operational energy security needs, which is key to the success of our Nation.

What do we need? For our industry to move forward, we've got the facility; we've got the supply feedstock agreements; we've got the market, we think, with the commitments from the military. However, we don't have the investment. We need $300 million to build, start, commission and operate this facility. And to get that type of investment, we need commitment from a significant off-take.

We think the most important market off-take that is not solely dependent, but an important key first start, is the Department of Defense. With our plans, what we need to succeed, the military to increase its operational security, collectively we need the ability to obtain long-term contractual commitments from the military to purchase these fuels. This would justify to the investment community the significant capital to build this facility. The Department of Defense is ideally situated to purchase these fuels, both from the region, as well as all throughout the country. All other aspects of our industry are in place except for this purchase commitment.

To point out a few reasons why we think the Pacific Northwest is critical, the combined use of military commercial aviation fuel in the Pacific Northwest is over 100 million gallons a year, annually.

We have advised the Pentagon that we can supply up to 80 million gallons a year of renewable drop-in fuels by 2014, with no technology risk, using existing feedstocks that exist now today for the same things that we use for biodiesel production, as well as providing an initial key market for crops like algae, crops like camelina as they scale and grow.

We are working with USDA on a number of different fronts. We have submitted for a loan application under the 9003 program, and we do think that continued support of USDA's bioenergy programs are critical for the biofuel industry to move forward. They're very important. We use those programs now in the biodiesel industry, and for the renewable jet fuel industry in the future.

We think the best path forward for advanced biofuel such as renewable jet fuel would be enable the Department of Defense to enter up to 15-year contract opportunities to purchase these fuels.

The commitment of the Defense Department will drive the entire renewable fuel—aviation fuel industry, which will help meet the demand on the civilian side, as well. As Navy Security Ray Mabus has stated, biofuels are a huge asset in providing the U.S. Navy, not just in promoting our Nation's defense, but in saving the lives of our soldiers.

I'd like to recommend to the members of the Committee a September 27 report from the Center for a New American Security entitled, Fueling the Future, Preparing the Department of Defense for a Post-Petroleum Era. This report outlines in great detail how the military should transition to a future that does not depend on petroleum, which currently supplies over 70 percent of the Department of Defense's energy needs.

By facilitating the development of advanced biofuels for the military with purchase commitments, this industry can become a major source of energy for both the military and civilian fleets. It also ensures the Department of Defense can continue to meet its mandate to protect the energies—protect the Nation's energy by reducing its dependence on petroleum supplies from Venezuela, from Iraq, from other places that we are certainly concerned about.

I would like to submit this copy of the report for the record, as well.

[The information referred to is contained in the Appendix.]

I want to thank you, Madam Chair, and Senator Murray, for sponsoring legislation that enabled the Department of Defense to enter into 15-year contracts to purchase renewable fuels. I recognize any contract will be competitively bid, but I am confident that Imperium is well-situated to prevail.

In closing, I appreciate the Committee's focus on the important issue. I certainly think advanced biofuels such as renewable jet fuel are a key to a cleaner, more sustainable, and most importantly, a more secure aviation industry. Like all forms of energy production that exist in America today, renewable aviation fuels needs a long-term, stable set of Federal policies that support this industry, and are critical to commercializing this fuel. This will assist in fulfilling the U.S. military's renewable fuel needs and promoting our Nation's security.

I do strongly believe that the success of biofuels for aviation will provide a tremendous benefit for generations of all Americans to come.

Thank you for your time and the opportunity to speak today.

[The prepared statement of Mr. Plaza follows:]

PREPARED STATEMENT OF JOHN PLAZA, PRESIDENT AND CEO, IMPERIUM RENEWABLES, INC.

Madam Chairwoman, my name is John Plaza. I am the President and CEO of Imperium Renewables, headquartered in Seattle, Washington. I very much appreciate the opportunity to appear before the Subcommittee today on the important issue of renewable aviation fuels. I also want to thank you, Senator Cantwell, for your continued leadership on renewable fuels.

Imperium Renewables owns and operates one of the largest biodiesel facilities in the world, located in the rural community of Grays Harbor, Washington. We have invested over $90,000,000 in our state-of-the-art biodiesel production facility at this site. We currently employ 42 people, five of whom are veterans who, after having served in Iraq and Afghanistan, have returned home to find family wage jobs in their community. Since 2007, our company has provided over $125,000,000 of direct economic benefit inclusive of payroll, taxes and revenue to other small businesses around Washington State.

We are planning to construct an advanced biofuel facility that will produce renewable jet fuel adjacent to our existing site in Grays Harbor. This new facility will create over 300 construction jobs during the first three years, and increase our workforce by an additional 50 permanent employees. With the construction and operation of this additional facility, over $250,000,000 will be invested by Imperium into Washington State during the construction phase, and once in operation, we will provide over $20,000,000 of annual direct economic benefit to the state. We are committed to the Grays Harbor community and believe that the Pacific Northwest, along with the entire nation, will benefit economically and environmentally from the development and use of renewable jet fuels.

Developing and deploying renewable aviation fuels has long been a dream of mine. Before founding Imperium Renewables, I was an airline pilot for over 20 years, having flown everything from small bush airplanes in Alaska to Boeing 747s around the world. I have a deep understanding of the critical importance of fuel quality, security and price for the aviation industry. Imperium Renewables was the first commercial producer of renewable aviation fuel. We produced the bio jet fuel that was used in a 2008 demonstration flight by a Boeing 747 operated by Virgin Atlantic Airlines. With additional successful demonstration flights since 2008, along with the recent approval by the American Society for Testing and Materials for renewable jet fuel to be used at a 50/50 blend, it is clear that renewable aviation fuels are ready for commercialization now. The market potential for these advanced biofuels is significant in our region. The combined use by military and commercial aviation in the Northwest creates more than 800 million gallons of jet fuel demand annually.

At Imperium, we have an aggressive plan that will enable us to produce significant amounts of renewable jet fuel by 2014. Importantly, this plan is contingent on obtaining long-term contractual commitments to purchase the fuel in order to justify the significant capital investment of over $250,000,000 required to build this new facility. The Department of Defense is ideally situated to purchase these fuels, which will facilitate the ability to raise the capital required to build advanced biofuel facilities. We have been in discussions with the Department of Defense concerning supplying multiple renewable jet fuel solutions to meet the military's needs in the Pacific Northwest region. We have advised the Pentagon that Imperium can supply up to 80 million gallons of "drop-in" renewable fuel by 2014, with no technology risk, while using existing feedstocks that are commercially available now and future dedicated energy crops that are in development.

The best path forward for advanced biofuels such as renewable jet fuel would be to enable the Defense Department to enter into 15-year contracts for fuel supplies to meet the demands of its facilities in the Pacific Northwest and around the nation.

The commitment of the Defense Department to renewable fuels will drive the entire renewable aviation fuel industry for the nation. It will also provide the Department of Defense a critical and important path forward in obtaining operational security of energy supplies right here at home. By its commitment to purchase renewable aviation fuels, all branches of our Nation's military can have secure regional sources of "drop-in" renewable fuels to better facilitate national security, as well as providing economic development and job creation for America, in America. As Navy Secretary Ray Mabus has stated, biofuels are a huge asset in providing the U.S. Navy operational security, not just in promoting our Nation's defense, but in saving the lives of our soldiers.

I would commend to the members of the Committee a September 27, 2010 report from the Center for a New American Security entitled "Fueling the Future Force: Preparing the Department of Defense for a Post-Petroleum Era." This report outlines in great detail how the military should transition to a future that does not depend on petroleum, which currently supplies over 70 percent of the Department of Defense's energy needs. By facilitating the development of advanced biofuels for the military with innovative technologies and fuel source diversification, along with improved efficiency, the Department of Defense can deal with future instability in petroleum supplies, reduce volatility of price spikes, and ensure it can continue to meet its mandate to protect the nation's security. I would like to submit a copy of this report for the record.

I want to thank you, Madam Chair, and Senator Murray, for sponsoring legislation to enable the Department of Defense to enter into 15-year contracts to purchase renewable fuels. I recognize that any contracts will be competitively bid, and am confident that Imperium is well situated to prevail.

In closing, I appreciate the Committee's focus on this important issue. Advanced biofuels such as renewable jet fuels are the key to a cleaner, more sustainable, more secure aviation industry. Like all forms of energy production that exist in America today, renewable aviation fuels need stable long-term federal policies that support this industry and are critical to commercializing the fuel. This will assist in fulfilling the U.S. military's renewable fuel needs, and promoting our Nation's security. The success of biofuels for aviation will provide tremendous benefits for generations of Americans in the future.

Senator CANTWELL. Well, thank you, Mr. Plaza.

And I thank all the panelists for their testimony.

I'm going to start with a question in general about the feedstock. We've heard from the first panel and, obviously, Ms. Canales about what USDA is doing.

Can you comment on the variety of feedstocks that could be used in this? Obviously, there's a lot going on in the Pacific Northwest, but there's opportunity all across the country for this product, I guess from the input to produce uniform output, I guess, is the best way to say it. Is that correct? And could you comment on what those opportunities are?

Ms. CANALES. Senator, thank you, and I can speak to some of the feedstocks that USDA is currently looking at. These are, again, not the completely inclusive list. But, oilseeds in the Midwest and Great Plains, and Pacific Northwest; poplar and pine, found in the American Southeast; sorghum, found in Texas and Oklahoma, and the lower Midwest; sugarcane, found in Florida and the Southeast; switchgrass in the Midwest and Southeast; soon to be announced will be others pertaining to Douglas fir, alder, hemlock, and eucalyptus; perennial and prairie grasses other than switchgrass; and then, of course, algae.

Senator CANTWELL. And so, are these in the end product an interchangeable source? Or are we talking about picking winners and losers here?

Ms. CANALES. Those are all prospects. Those are all opportunities for feedstock. So, this is not in any type of order of priority.

Mr. PLAZA. So, from Imperium's perspective, the technology that's ready for scale and commercialization now is using lipids, such as existing oilseed crops, future oilseed crops, like algae and camelina. We specifically have invested, as you heard from Secretary Yonkers, in the technology and development of converting alcohols into jet fuel, and have partnered with Pacific Northwest National Labs and the Tri Cities to develop a technology that will allow us to use feedstocks like municipal solid waste and wood residue.

Just using Seattle as an example, we produce enough trash every day to make 200 million gallons of jet fuel if our technologies can scale. So, the importance of, I think, this industry cannot be dependent on any one feedstock or any one technology, but dependent on the finished product. It's important that the end result be approval of these products, and acceptance of these products, and long-term policies that support multiple pathways.

And we actually believe longer-term, as Secretary Yonkers mentioned, that alcohol-to-jet fuel is a better pathway in the long run than lipids.

Senator CANTWELL. Mr. Altman, did you want to comment on that?

Mr. ALTMAN. Yes. All of these feedstocks are applicable, and from the CAAFI point of view, we are neutral to process. From the certification point of view, we have chosen to group these together in processes. So, camelina, algae, renewable jet has been explained to you, alcohol's process can come from anything from woods to giant Miscanthus, which is prominent in the Southeast. It's really a matter of what grows locally, and what can be harvested and produced economically.

From our point of view, we're qualifying all of them. And as Mr. Yonkers and others have identified, we should have a full family, through alcohol-to-jet, through synthetic biology, using sugars, especially those from cellulosic sources and through pyrolysis. I would expect all of these to be qualified by 2015.

Senator CANTWELL. And so, that is what your fuel readiness level is, and is identifying as an industry, what fuels are at those levels?

Mr. ALTMAN. Yes. What fuel readiness level does is, it uses the accepted Defense Department gated risk management approach to manage how we bring along technology from the very, earliest research base through certification, and then through production. And this is something that has been globally accepted by ICAO, and we use this to track process.

CAFFI/FAA also have with USDA a variation of that called feedstocks readiness. We've taken aerospace techniques for project management, and are applying that with the research centers of USDA to mature feedstocks on a similar scale, and to communicate their status, as we do with the fuels qualification process for aircraft use.

Senator CANTWELL. Thank you.

I just want to try to get to this point, if I could. And maybe our panel is not the panel to ask this question. But, I'm talking about, now we're talking about a completed system here. And does, you're talking, you know, very near-term when members can see 2016 or something of that nature. But, you're talking about people being able to contribute a variety of feedstocks to produce one fuel source.

So, we're not talking about one of these winning as a concept. We're talking about each region producing fuel from a feedstocks that all goes into an aviation fuel. Is that correct? So that the scalability here is an issue of the, I guess, the, I don't want to call it magic. But, the scientific process of getting an alcohol-based product out of a variety of different fuel sources, and——

Mr. ALTMAN. That's absolutely correct.

Senator CANTWELL.—and then, still being able to supply that on a uniform basis across the United States.

Mr. ALTMAN. Absolutely correct. We refer to this concept as something other than the silver bullet. We refer to it as silver buckshot——

Senator CANTWELL. OK.

Mr. ALTMAN.—and that means, the buckshot applies to everyplace locally. As I mentioned, we ourselves are working with some 20 different states, all of whom have different approaches to the process. And it's going extremely well. We just need a continuation of the programs which USDA has and which DOE has to allow these processes and feedstocks to mature.

You know, one thing I've mentioned and just second to what Tom Todaro was telling you, is if you look at the history of food crops, like corn, and look back around 50 years, you'll see a 400 percent gain in yield per acre. Price is market-driven. Yield is something that we can work on, and work on together. And that's what we're doing with USDA.

Senator CANTWELL. Well, I think the buckshot analogy is appropriate. And I think to Mr. Plaza and Mr. Todaro's point is that then this becomes a very, very viable rural economic development strategy for the U.S. Because if every region is then providing feed source to an eventual end product, you know, everybody is, you know, everybody is contributing to the code of the operating system, if you will. And I think that's very, very positive.

I don't know, Ms. Canales, how you're dealing with that in your choices at USDA. But I'm assuming that you're looking to get regional viability.

Ms. CANALES. Yes, ma'am. Indeed. And, thank you, Senator. And what I would say to you, though, I've been onboard now almost two and a half years in this role, is that I have been promoting geographic diversity in the role of USDA rural development in my agency, and in utilizing the funds that have been made available by the Congress to incite, and to be able to incentivize these types of projects.

What we're seeing already right now is a complete diversification. Certainly, regarding 9003, as I mentioned to you, you know, there are 10 projects that we're looking at individually, and each one is separate. And they're coming from different parts of the country. There were five prior to that that we've already issued a conditional commitment to. And they, too, are in different parts of the United States.

And then, I should also say, within our other programs, Rural Energy for America program, that has become completely diverse. And that's notable, because that's getting initial momentum at the local level—businesses engage; the producers engage. And that has become a true national program.

Senator CANTWELL. Mr. Plaza, did you want to add to that?

Mr. PLAZA. Yes. I think, going back to a question a moment ago. The key, I think, for the industry is to focus on the end product being consistent, being molecularly identical to petroleum, or what's called commonly now drop-in fuels. That allows multiple production technologies, multiple feedstocks, and a, definitely a regional platform.

One of the things that we communicate to the Department of Defense as they develop strategies for renewable fuels is, in a long-term purchase commitment, don't pick a technology. Pick an ASTM-approved fuel that meets the, either the emissions under 526, which is an important category, the price, or a number of other important aspects, and allow the producer to develop a platform that provides that. And in our case, it's a multitude of platforms—one using the existing lipids that are available now, and then developing additional technologies that would make an equivalent drop-in replacement.

As you know, in the Pacific Northwest we have huge amounts of wood residue that go left unused, or have little to no value. That's the type of feedstocks that we're developing technologies to access, along with municipal solid waste, agricultural wastes. That works in the Southeast, but maybe it doesn't work in other parts. So, there'll be a combination of all of these. The key component is, we all make the same product, through various technologies.

Senator CANTWELL. And so, where do we think we are with woody biomass or algae as a source?

Mr. PLAZA. Well, I'll speak to woody biomass, and maybe let the others speak to algae, because they're probably more familiar.

I think woody biomass is quickly becoming a very viable feedstock. Whether it's, you know, 5 years away or 10 is still up in the air. I think it's important that we see more research and development from DOE; we see support in the funds provided to USDA, who are supportive of woody residues as part of their agricultural process; and we see long-term purchase commitments for these fuels that allow companies like Imperium to invest and, in fact, invest more to accelerate the development of that.

Senator CANTWELL. Mr. Altman or Ms. Canales?

Mr. ALTMAN. Well, let me deal with the algae part for you. We are very supportive of Sapphire. They're a stakeholder member as is Helia and a number of different companies.

There are two different approaches. One is a process used by a company called Solazyme, which actually uses algae as a catalyst or an enzyme to process fuel that is actually a HEFA/HRJ fuel. That process is commercial, and it's commercially viable now.

The issue with open pond and closed pond systems is to get the cost and economic benefit under control. Typically, there is significant requirement for energy use to extract water from algae. And we are tracking very closely with those companies to help make sure that within a period of 5 to 10 years we're able to get competitive costs from algae.

Ms. CANALES. I would just speak to Sapphire Energy as being one of the projects that we're so keen on, because of the fact that it pertains to exactly what we're trying to do here, is to develop an alternative source for aviation fuel. And Sapphire is online. You know, what we deal with within my agency, really, is the financing. And then we work with NREL and—National Renewable Energy Laboratory, as you're familiar with—to review the capacity, the technology, the, is this going to work? And so, we combine all of that. From our end, it's about trying to get these projects financed, and utilization of the, our loan guarantee program.

Senator CANTWELL. And does the ASMT have the capacity to, if you would bring together such a standard, I mean, obviously, they're investigated in the testing and approval of this. But, if you, if, one element of this is the standardization. Are they equipped to guide this next phase, Mr. Altman, in coming up with what that lipid standard or——

Mr. ALTMAN. Absolutely. And the thing that has happened in the last 5 years has been ASTM took a process that used to take 10 years to qualify fuel for one process from one factory, to approve fuel within a period of 3 years. We've been able to qualify both HRJ/HEFA and Fischer-Tropsch in 3 years. So we've learned how to manage the qualification process.

The issue that we're dealing with now is that we have three processes running in parallel toward qualification. When I mention the requirement for research funds, it is noted that the Committee was kind enough to bring the advanced biofuels research effort up to snuff at FAA for the one-year funding to pursue that potential. But when we look at having to do three processes in parallel, yes, ASTM has the capacity, but it's, requires that we progress through the FRL levels to make that happen. And it's a much more complex process when you're doing three than it is when you're doing one. But can we do it by 2013, 2015? The answer is, yes, we have that capacity.

Senator CANTWELL. So, they will end up approving three different processes? Or do you think a standard will——

Mr. ALTMAN. Well, we're—you're asking a very good question, because that's what the Committee is working on right now, is to establish how many different processes do they actually have to have. I mentioned three, because one's biological, from a synthetic biology process. The other is catalytic, to get the alcohols. And the third is pyrolysis. So, my mind categorizes them that way.

Whether the next time that the ASTM committee meets, they'll have three processes or one process is really up to the technical committee to establish that. But I think we have to be prepared to look at multiple pathways at this point, and not count on, as we did in the past, on having a single pathway, like HRJ, HEFA or Fischer-Tropsch.

Mr. PLAZA. If I could add a couple things.

First of all, with respect to funding agencies to do the work to develop these new requirements, I think AFRL, the Air Force Research Laboratory, is critical. They do a lot of work with CAAFI, with both the civilian and military with the OEMs. Their funding has been challenged, obviously, with many other areas. But the AFRL group is a really important group out of Wright Patterson to continue to get funding.

I do want to step back and talk about where we are now, though, and point out that there are billions of gallons of lipids available to make this HRJ product that we and others would like to make. So, it's important that we recognize, there is significant quantities of oilseed crops that exist that we use for biodiesel today that don't interrupt food supply, have great life cycle emissions, tremendously valuable—and priced economically. And divergent to the previous panel, there is some significant speculation in that market as well. Equally damaging to the ability for biodiesel to be competitive against petroleum, and any future biofuels that would use commodity-based index pricing for the feedstock is the speculation that occurs there. So, just like petroleum, that occurs in grain products. It's much of what drives grain product pricing.

So, I think the more that we can broaden that commodity speculation concern against all things that folks—because if you kick them out of petroleum, they're going to go into grains. And that's going to cause even a bigger problem.

Senator CANTWELL. Well, I've certainly heard from a variety of commodity producers about the concern there. So, obviously, we are working very hard to try to commence the CFTC to broaden their efforts.

And then, just one last question, because I know we're getting close to the noon hour here.

And, Mr. Plaza, could you just give us an idea of what kind of scale we need to have for producing aviation fuel? What level of facility?

Mr. PLAZA. Well, it's an interesting point. I think that the military's commitment for the U.S. Navy, for example, 366 million gallons of renewable jet fuel by 2016, the U.S. Air Force's commitment, which is quite significant for alternative fuels, which is obviously more than just biofuel. We belief that the appropriate scale and the application we submitted to USDA is for 100-million-gallon-a-year capacity, which puts out about 80 million gallons of renewable jet fuel.

The market in the Pacific Northwest is certainly that size. The feedstock is available now from traditional oilseed crops used in biodiesel. It's the appropriate size and scale for where we are today. It certainly doesn't amount to a meaningful amount to displace significant amounts of petroleum. But I think it's a great commercial start to get all of these strategies and feedstocks to the table. And I think there's opportunity for us to see those replicated five to ten times across the country. And so, I think that's the appropriate scale. It can be done smaller. But like with anything, the smaller you go, the more expensive it is on a per gallon basis scale.

Senator CANTWELL. So, you're saying—and I know this is a hard question to answer, just as Mr. Todaro was saying, you know, tell me whether I'm viable or not. And he said, obviously it depends on what the price of oil is. And we've seen a lot happening in that marketplace. And obviously, we want something that's far more predictable for the future.

But, you're saying 100 million? Is that what you are saying is the single——

Mr. PLAZA. 100 million gallons a year scale is an appropriate scale to bring the costs to commercial levels. For example, the Defense Logistics Agency just had an RFPL for about 450,000 gallons of a combination of renewable diesel and renewable jet. We submitted a response to that that was a little bit of a unique idea. But the pricing we put in was for a smaller scale unit.

But the prices with today's policy, which is an important component—taking into account the renewable fuel standard, as you heard from Mr. Glover, the long-term extension of biodiesel credit which applies to renewable jet fuel—inclusive of those policies we're looking at commercial fuel prices in the $5 to $6 a gallon range, which is still more than petroleum. But for a nascent industry at commercial scale, I think it's significant. I think we can drastically reduce that with new technologies.

Senator CANTWELL. Mr. Altman, is your association looking at these issues of scalability?

Mr. ALTMAN. We're obviously very concerned with that. And I would agree with your remarks.

I think the important thing is, when you get that 100 million gallon facility going, we're looking at a very large opportunity, I think, for the investment community to say, yes, we can participate in this. Potentially oil companies, also other individual investors will jump in with investments. And what's really holding back the industry right now is investment.

So, to have that 100 million gallon facility, that will be a very important element for us in terms of triggering that next round of significant investments. That is what USDA has been trying to achieve.

Senator CANTWELL. And, this may again be a question, maybe, not for this panel. But in general, I think the Europeans have a jet fuel SPRO, do they not? So, they are basically reserving jet fuel as a way to help build stability. And I, if we're looking at our current situation here, another way to look at this is just to say, you know, that you're going to produce an, you're going to produce green jet fuel as a way to even help mitigate price in the future. If we're already doing a SPRO, why not help stabilize prices by having an alternative source that you could be producing?

Mr. ALTMAN. That is a good question for me to answer, because I was part of the swathee European SWAFEA biofuels initiative, which is the path that the EU pursued.

Until a year ago, they had effectively no jet alternative fuel program. They really ran a program to establish if ETS was sufficient to achieve their goals. And quite frankly, it held back their jet fuels program.

They now have a program, and it's really sprung up in three different places—in Germany, in France and in Spain. CAFFI had an exhibition at the Paris Air Show. We were privileged to have Secretary Vilsack there, Secretary LaHood. And at that particular show we had each of those international initiatives come into play. But, I can tell you, what USDA is doing, what DOE is doing, what FAA is doing are the envy of all the groups in Europe. And they don't really have a process.

That's why 80 percent of the fuel projects around the world in biofuels for aviation come from U.S. companies. It's a great export opportunity for us.

Senator CANTWELL. I think we are exporting. So——

Mr. ALTMAN. Yes.

Senator CANTWELL. Anyway. Well, thank you very much to the panel. Appreciate everybody's testimony today.

We'll leave the record open. If my colleagues have questions, we hope you'll respond to those and submit those for the record.

Senator CANTWELL. But, thank you for pioneering in this very important area of green aviation fuel.

This hearing is adjourned.
[Whereupon, at 12:02 p.m., the hearing was adjourned.]

APPENDIX

PREPARED STATEMENT OF JIM REKOSKE, VICE PRESIDENT/GENERAL MANAGER, RENEWABLE ENERGY AND CHEMICALS, HONEYWELL/UOP

Senator Cantwell, Senator Thune, and distinguished members of the Subcommittee; thank you for allowing me to testify today on behalf of Honeywell UOP's Renewable Energy and Chemicals business. As Vice President and General Manager of the business unit, I am very excited to submit the following progress report to the Aviation Operations, Safety, and Security Subcommittee.

In 2007, Honeywell UOP leveraged its nearly 100 years in refining technology and process expertise to form its Renewable Energy and Chemicals business. Tasked with providing new and adapted technology for processing renewable energy sources in petroleum refineries, the business unit has grown considerably in the past 5 years, and is a clear leader in the drop-in renewable transportation fuels market.

Since its inception, our team has successfully developed and tested multiple forms of transportation fuel including renewable diesel, renewable gasoline, and renewable jet fuel made from a variety of sustainable feedstocks available around the globe. The robust processes that have enabled our team to so rapidly jump over the technological hurdles that we have faced have also enabled our supply chain partners, and the entire renewable fuel segment to grow to a point where we are on the precipice of providing real volumes to end use customers.

The cornerstone of these achievements has been the successful production and test of Honeywell Green Jet Fuel.™ From our initial contract with the Defense Advanced Research Projects Agency (DARPA) for the development of jet fuel from renewable sources, we have been a willing partner of both private industry and the Department of Defense. Our Green Jet Fuel has been proven in 17 military and commercial applications to date, and the more than 700,000 gallons we have produced have performed exceptionally. Notable features include:

- True feedstock flexibility as our fuels have been made using a variety of purpose grown inedible crops, algal oil, and animal fats
- Performance at or above all flight specifications, including running with a higher energy density in flight than petroleum fuel
- A reduction in net carbon emissions up to 85 percent lower than petroleum jet fuel
- Powerful enough to fuel the first renewable supersonic flight of the Navy F/A 18 Green Hornet
- Precise enough to power the Air Force's premier demonstration team, the Thunderbirds, in acrobatic performances
- Robust enough to fuel the world's first renewable transatlantic flight of a Gulfstream G450

These achievements are notable as we are at a critical point in the developmental history of the renewable fuels industry. We owe the members of this committee and your colleagues in the defense community a debt of gratitude for your partnership and support over the last 5 years.

Honeywell refining technology has been used to produce 100 percent of the fuel used by the Department of Defense in seven demonstrations conducted on various Army, Navy, and Air Force platforms including final certification of the Air Force C–17 and F–16 aircraft. Fuel made from Honeywell technology has also been used in ten commercial demonstration flights to date. These flights, along with collaboration with the FAA, ASTM International, and other industry leaders, helped to establish the proper specifications for aviation fuel made from natural oils and fats and resulted in the recent approval of this product by ASTM International for commercial flight.

Thanks to the efforts of the Federal Government, we are one step closer to realizing commercialization of these products for both defense and domestic use. However, we at Honeywell (like everyone in this room) are not satisfied with the progress made to date. Only when we are refining and selling volumes measured in millions of barrels, rather than thousands of gallons, will we be pleased with the state of the industry.

It is for this reason that we are writing today. We urge the members of this committee to support and heed the advice of those testifying. It is because of the efforts of those on the panel that we have taken such a strong position. It is worth mentioning instrumental partnerships:

- *Leading aircraft manufacturer, Boeing* has been a significant partner to Honeywell both for the testing of feedstock sustainability and in multiple in-flight tests. A leader in the aviation field, their cooperation and input has been critical throughout our development process. Boeing's commitment to sustainable fuel is proof that industry leaders are dedicated to growing this field. We will continue our partnership with both large and small industry players, on programs that promote commercial use of biofuels and increased energy independence as well as the creation of "green" jobs to support a new biofuels infrastructure.
- *Future customer, The United States Air Force* has conducted five demonstration flights with Honeywell Green Jet and has been a key partner in helping our team learn how best to design fuels for defense aviation.
- *Feedstock provider, Sustainable Oils* has played an important role in the growth and acceptance of Camelina as a feedstock for renewable jet fuel. Honeywell and Sustainable Oils have collaborated on every demonstration of camelina-based renewable jet fuel to date. Acting as the harvester and processor of camelina oil, Sustainable Oils has provided Honeywell with high volumes of oil, which we have successfully converted into jet fuel. Without sufficient feedstock, Honeywell Green Jet Fuel cannot be a reality, and we believe that in the near term, camelina offers the promise of providing real volumes for commercial use. It is thanks to companies like Sustainable Oils that we have been able to perfect our refining process, and deliver needed gallons to both private industry and the defense community.
- Finally, *regulatory and technological enabler, the Federal Aviation Administration* (FAA) has funded the development of mature technology for Fuel Burn Reduction and test aviation biofuels for use in the Gas Turbine Engines from our aerospace division. This is enabling emissions benefits and cost savings for both the hardware of flight and the fuel used to power it.

I mention every company testifying because nearly the entire supply chain is represented in today's pane. We at Honeywell are working across the spectrum for solutions to our Nation's energy needs, believing that only through an honest appraisal of how we consume energy, can we make a real impact, and rise to the energy demands of the future while maintaining sustainability.

Thank you for taking the time to hold this hearing and consider our testimony. We stand ready to push this industry toward commercialization, and will continue to act as the turnkey solution provider for the multitude of renewable fuels companies that will meet the demand of the aviation industry and beyond.

RESPONSE TO WRITTEN QUESTION SUBMITTED BY HON. JOHN D. ROCKEFELLER IV TO DR. LOURDES MAURICE

Question. What are the primary activities the FAA supports to foster the development of alternative fuels?

Answer. The FAA is engaged in a number of activities. The major ones include:

a. The FAA has the responsibility to make sure that any aircraft, aircraft engine or part, or fuel that is used in aviation is safe and performs to set standards. The FAA does not directly approve fuel but rather approves aircraft to operate with fuel that meet specifications, such as those set by ASTM International. The FAA is very engaged in participating in the process to set specifications—FAA staff participates in ASTM International and FAA has sponsored testing to support the ASTM international process. The FAA works very closely with our sister departments and agencies as well as industry in these endeavors.

b. Through CLEEN, the FAA is sponsoring efforts—in collaboration with industry—to test fuels and evaluate fuels. Last year, FAA received additional resources for alternative fuels, and is pursuing efforts to conduct engine durability tests with alternative fuels, and to perform key testing to support qualification and certification of novel jet biofuels from alcohols, pyrolysis, and other processes.

c. FAA is engaged in environmental evaluations—from direct engine measurements to analyses to establish life cycle emissions and sustainability criteria. That work is primarily performed through the Partnership for AiR Transportation Noise and Emissions Reduction (PARTNER) Center of Excellence as well as the John A. Volpe National Transportation Systems Center.

RESPONSE TO WRITTEN QUESTIONS SUBMITTED BY HON. MARIA CANTWELL TO DR. LOURDES MAURICE

Question 1. Dr. Maurice, the Future of Aviation Advisory Committee Report to Secretary LaHood considers success if "approximately 5 percent of aviation jet fuel could come from sustainable low-carbon lifecycle sources by 2020. These new fuels could reach majority status by 2050." Do you believe the 5-percent goal is achievable?

Answer. The goal of having 5 percent of aviation fuel comes from alternative fuels by 2020 translates to roughly a bit over a billion gallons. The FAA has set a target for use of 1 billion gallons of alternative aviation fuel by 2018, which is consistent with the 5 percent and which the FAA believes is achievable.

Question 1a. Do you believe it is ambitious enough target?

Answer. The goal is aspirational and very ambitious. It requires that the industry grow from making thousands of gallons to a billion gallons in 7 years. The good news is that the fuels are qualified for use and we know how to make them. Moreover, we are working hard to qualify additional classes of fuels.

Question 1b. What do you see as the major challenges for alternative fuels to reach majority status prior to 2050?

Answer. The challenges are building the infrastructure to produce the fuels, including feedstock availability. Also, we must make a careful examination of sustainability; that is, we have to make sure there are no unintended consequences. However, the FAA is confident that with the will, which includes resources, our Nation and our industry are up to the task.

Question 2. Dr. Maurice, if the overall federal strategy is to have as many different feedstock and process pathways for creating alternative fuels as possible, having ASTM certifying fuel specification using different processes are critical. Which processes do you see as being next in the queue for ASTM to approve?

Answer. Jet fuels from other advanced processes such as pyrolysis, alcohol oligomerization, and advanced fermentation are being investigated. Fuel samples are being evaluated and the testing to support ASTM approval in underway.

Question 2a. In the Senate-passed FAA bill, Senator Warner and I created a grant program for conducting research in the use of alternative fuels as well as a Center of Excellence for Alternative Jet-Fuel Research in Civil Aircraft. Assuming the section remains in conference report, do you believe that grant program and the Center can be utilized to collect the data necessary to speed up the standards process for certifying fuels specification made with new processes?

Answer. Such a Center of Excellence could serve this purpose, along with efforts by industry and government labs. We note, however, that Centers of Excellence are generally university-led and the potential university participants do not generally have the expertise and infrastructure for collecting data to support fuel specification approvals. The Center of Excellence might be more valuable in exploring sustainability criteria and to identifying new classes of fuels. We will, of course, structure the Center to best advance national alternative aviation fuels goals and in accordance with any legislation.

Question 3. Dr. Maurice, what is the status of using biofuels for general aviation aircraft?

Answer. Separate from our work with jet fuel, the FAA is partnering with industry to investigate unleaded alternative fuels to replace the current avgas. However, avgas presents a more difficult technical hurdle to identify a "drop-in" replacement fuel, and the limited resources of the GA community make this an even more challenging initiative.

General aviation faces mounting environmental pressure to curtail its use of lead-containing aviation gasoline (avgas) in piston engine aircraft. Lead is a known toxic substance, but it provides performance benefits that aircraft piston engines rely on. The FAA is committed to ensuring that the approximately 190,000 small aircraft powered by piston engines have a safe fuel to perform their wide variety of important roles.

FAA continues to support research on unleaded avgas, including bio derived avgas alternatives, at the FAA's William J. Hughes Technical Center in Atlantic City, New Jersey. FAA recently expanded its role by establishing an Aviation Rulemaking Committee (ARC) to develop a go-forward strategy and plan to help industry develop and deploy an unleaded avgas. The ARC is comprised of key stakeholders from the General Aviation and fuel industry. It is currently in progress and expects to issue a recommendation by early calendar year 2012.

Earlier this year, the FAA and industry stakeholders led the effort at ASTM International to approve a grade of avgas with 15 percent less lead. This new avgas, called 100VLL (for "very low lead"), can be used on all current aircraft and is intended as an interim solution to a completely lead-free fuel.

Question 4. Is there any storage issues associated with aviation biofuels? Do you expect aviation biofuels to degrade if they are stored for appreciable periods of time?

Answer. Unlike biodiesel, aviation biofuels are essentially the same chemical composition as petroleum-derived jet fuel, so they are considered "drop-in" fuels. These drop-in fuels do not have any special storage or handling requirements and they can be freely co-mingled with the jet fuel currently in use today. However, we are being cautious and conducting studies to make sure there are no issues with long term, prolonged use of aviation jet biofuels.

RESPONSE TO WRITTEN QUESTION SUBMITTED BY HON. JOHN D. ROCKEFELLER IV TO BILLY M. GLOVER

Question. Can you discuss broadly how the development of alternative fuels creates job growth across the supply chain?

Answer. Focusing the power of U.S. agriculture on the production of alternative fuels creates jobs in rural America (building and operating processing facilities) and increases farm income. Additional markets are created for growers, bolstering farm income; jobs are created to transport pre-processed raw biomass; construction jobs are added to add fuel processing and storage facilities; fuel processing and distribution jobs are added as new processing capability comes on-line; and finally, the new fuel strengthens the sustainability of both commercial and military aviation. Commercial aviation drives $1.2 trillion in annual economic activity and 11 million well-paying American jobs (source: Air Transport Association of America.)

RESPONSE TO WRITTEN QUESTIONS SUBMITTED BY HON. MARIA CANTWELL TO BILLY M. GLOVER

Question 1. Mr. Glover, you view Boeing's role as facilitating multiple feedstocks, multiple processing methods, and multiple supply chains globally. My understanding is that the company continues to be active in the ASTM standards development process. According the Dr. Maurice's testimony, the next few candidate aviation fuels will be far enough along for consideration by the standards organization in 2013 timeframe. Are there things the Federal Government do to accelerate the standards process?

Answer. Due to the increased attention over the last few years, we now have a much improved jet fuels approval process. However, scaling up new fuel processing methods to produce sufficient quantities for testing, and actually producing fuel and running the engine-related tests remain challenges. Lack of funding for these activities is the pacing item in completion of jet fuels approvals. Federal funding for processing methods development and fuels testing would greatly assist the standards approval process. Use of the unique facilities of the Department of Defense and National Aeronautics and Space Administration would also speed approvals.

Question 2. Mr. Glover, as you mentioned in your testimony, Sustainable Aviation Fuels Northwest made a number of recommendations* regarding the creation of commercially viable aviation biofuel supply chain in the four state region that includes Washington, Oregon, Idaho, and Montana. One idea is for legislation to extend the tax credit under Section 40A for producers of biodiesel, renewable diesel, and certain aviation fuels derived from biomass. Why would extending the tax credit make such a difference?

Answer. Extending the tax credit helps to make the business case for a nascent industry where every advantage is needed to get through the initial stages of a new start by fostering attractive market conditions for investment. Adding certainty to

*This report is available at *http://www.safnw.com/wp-content/uploads/2011/06/SAFN_2011Report.pdf.*

such a tax credit also is critical. Providing a tax credit for a definitive period of time provides certainty for businesses and financiers to make long-term investment decisions thereby speeding up the market entry for these fuels, which in turn, will bring greater economic development.

RESPONSE TO WRITTEN QUESTION SUBMITTED BY HON. JOHN D. ROCKEFELLER IV TO TOM TODARO

Question. Can you discuss broadly how the development of alternative fuels creates job growth across the supply chain?

Answer. The development of alternative fuels, particularly biomass-based fuels, creates jobs in numerous sectors, at various skill levels, in geographic regions throughout the United States.

- *Feedstock Supply* creates jobs in both rural and non-rural areas, including high tech research and development for higher yield, more sustainable varietals of energy crops; jobs in farming and related agriculture industries (*e.g.,* seed, water and fertilizer services); and jobs related to the collection, harvest, storage, transportation and delivery of biomass to biorefineries.
- *Fuel Production* creates jobs in the planning, design, engineering, construction, operation and maintenance of biorefineries.
- *Fuel Distribution* creates jobs in not only the transportation sectors (*e.g.,* truck drivers) but also in the planning, design, engineering, construction, operation and maintenance of alternative fuel pipelines and associated infrastructure. Alternative fuel distribution also creates jobs in retail fuel distribution.

RESPONSE TO WRITTEN QUESTION SUBMITTED BY HON. MARIA CANTWELL TO TOM TODARO

Question. Mr. Todaro, one of the ways you recommend that Congress can help reduce or remove some of these obstacles to widespread commercialization is to ensure that EPA makes a clear determination that camelina-based jet and renewable diesel fuels qualify under the existing Renewable Fuel Standard. Can you explain to this committee why qualifying for a Renewable Identification Number and receiving market-based credits is important?

Answer. Without a Renewable Identification Number, a domestic renewable fuel producer may not sell renewable fuel in the United States. Moreover, petroleum refiners and importers are not obligated to procure "RIN-less" fuel, as these entities may only comply with their annual volumetric obligations under the Renewable Fuel Standard (RFS2) by submitting valid RINs associated with volumes of renewable fuel. Camelina meets all necessary criteria to qualify as a form of renewable biomass under RFS2, but EPA has yet to certify a fuel pathway for fuel derived from this feedstock. The failure to certify camelina's eligibility as a form of renewable biomass under RFS2 in the near future could impose significant legal and commercial barriers to the production of camelina-based renewable fuel.

RESPONSE TO WRITTEN QUESTION SUBMITTED BY HON. JOHN D. ROCKEFELLER IV TO SHARON PINKERTON

Question. How do you see the development of biofuels fitting in with the airlines' goals of reducing emissions and limiting the impact of aviation on the environment?

Answer. Although the U.S. airlines contribute only about 2 percent of the Nation's carbon dioxide (CO_2) inventory, our airlines are committed to continuing their strong record of fuel efficiency and CO_2 emissions savings and are relentlessly pursuing an array of initiatives in this regard. At the core of these measures is the ATA carriers' commitment to technology, operational and infrastructure measures to continue our drive toward ever-greater fuel and CO_2-efficiency improvements. This includes tremendous airline investment in new aircraft, new aircraft engines, navigation aids and enhanced operational procedures. In addition, ATA and its airlines are dedicated to developing commercially viable, environmentally friendly alternative jet fuel, which could be a game-changer in terms of aviation's output of CO_2. To this end, as noted in my testimony, ATA is a founder and co-leader of the Commercial Aviation Alternative Fuels Initiative (CAAFI)®, a consortium of airlines, government, manufacturers, fuel suppliers, universities, airports and other stakeholders working to hasten the development and deployment of such fuels.

RESPONSE TO WRITTEN QUESTIONS SUBMITTED BY HON. MARIA CANTWELL TO SHARON PINKERTON

Question 1. Ms. Pinkerton, you heard Mr. Yonkers speak about the Air Force's ambitious plans for purchasing alternative jet fuels. I know the Navy has similar plans. Over the next few years, how important are the Department of Defense's purchasing decisions in shaping the market for alternative fuels for commercial aviation and building capacity for the different feedstock producers and processors?

Answer. As noted in my testimony, the aviation industry and would-be alternative jet-fuel suppliers are on the cusp of creating a viable alternative jet-fuel industry. But government support is needed in the near team to provide financial bridging and other tools necessary to help us get over the cusp. One area of support is having the military join the commercial airlines in sending the market signals necessary to give investors the confidence to invest in the aviation alternative fuels industry and to stimulate would-be producers to go forward. In this regard, the military's plan to purchase alternative aviation fuels is critical. However, as recognized in the Strategic Alliance between ATA and the Defense Logistics Agency, the procurement arm for the military, military demand alone is not enough of a market signal. As commercial aviation represents a much greater portion of demand, it is important that we continue to work together, as contemplated in our Strategic Alliance. (More detail on this alliance is available at ***http://www.airlines.org/News/Releases/*** *Pages/news_3-19-10.aspx*). Further, market signals are only one part of the equation. As I noted in my testimony, it is critical that existing federal programs that have been effective in supporting development and deployment of alternative aviation fuels be maintained and, if possible, expanded.

Question 2. For example, in this early stage of market development, would a DOD decision to enter into a long term contract to purchase biofuel or synthetic fuel, exclusively, give one fuel such a first mover advantage that the other, or yet to be approved fuel specification such as alcohol-to-jet, will never be able to get a toehold?

Answer. We do not see a determination by the military to enter into a long-term contract with any particular supplier as providing an obstacle to other suppliers or to promising alternatives. As noted, the military is only a small portion of U.S. demand for jet fuel—needing about as much per year as a mid-size commercial airline. Thus, it is difficult to envision it swinging the market. What is a greater worry is what will happen if the U.S. military were to not be able to pursue alternative fuels with vigor. Without the military joining commercial aviation in the pursuit of alternatives, the opportunity for a strong market signal would be reduced and the military would not be in line for much-needed supply. That is why ATA is on record as supporting legislation that would give the military the authority to enter into long-term contracts, which it does not now have.

RESPONSE TO WRITTEN QUESTION SUBMITTED BY HON. MARIA CANTWELL TO JUDITH CANALES

Question. Ms. Canales, in order to have a sustainable aviation biofuel system, it is imperative that sources of biomass be identified that will be reliably available. The non-food crops, such as camelina, are relatively primitive, requiring robust and ongoing genetics, breeding, agronomic studies so these crops can be optimized for various production locales. Such research and develop is historically conducted by America's land grant universities. What steps will USDA take to ensure that the necessary agricultural research and grower education is funded for the development of non-food crop as feedstocks for aviation biofuels?

Answer. USDA understands that support to our farmers and researchers is necessary to provide adequate non-food feedstocks for the domestic production of biofuels. To ensure that research, education, and extension efforts are coordinated, the USDA has followed the guidance of the President's Biofuels Interagency Working Group's *Growing America's Fuels* report, and developed a coordinated effort directed to help supply the agricultural and forest-based feedstocks that are needed to support commercial production of aviation and other advanced biofuels.

One particular effort has been the USDA establishment of five regional USDA Biomass Research Centers which were established between the Agricultural Research Service (ARS) and the U.S. Forest Service Research and Development (FS) utilizing the two agencies' nation-wide networks of scientists and facilities. The National Institute of Food and Agriculture (NIFA) Agricultural and Food Research Initiative (AFRI) is also offering large competitive grants for coordinated agricultural projects or CAPs that look to support regionally directed research and grower education through extension to deploy improved feedstocks and production systems that will produce aviation biofuels. NIFA currently plans to offer this competitive program again in Fiscal Year 2012. The Regional Centers and AFRI CAP projects have been designed to focus, coordinate, and accelerate the science and technology needed to incorporate feed stock production into existing agricultural and forest based systems.

Specific to the development of feedstocks for aviation fuel production, a network of ARS research facilities in the western U.S. in cooperation with NIFA supported land grant universities and industry partners are determining how to incorporate camelina and other oil seed crops that can be produced in rotation with cereal and other crops. Research is also being done with support from the Navy Office of Naval Research to determine where oil seeds can be produced on marginally productive and abandoned lands and least adversely impact food crop yields and existing commodity markets. Both extramural and intramural USDA research supports the genetic development of other high-performance dedicated biomass feedstocks and sustainable systems for their production. NIFA supports work on biofuels through several programs. For example, NIFA formula funds are made available to land grant universities, genetic and breeding research supports improvements to non-food crops for bioenergy and biofuels production, and the extension of this information to growers. These funds also support work on the production and management of these crops. Fundamental research for crop improvement, diseases, pests and the economics of feedstock crops is supported by the AFRI foundational competitive grants program.

The Biomass Research and Development Initiative competitive grants program supports research, development and demonstration projects that address three legislatively identified technical areas at the same time in each supported project. These areas include: (A) Feedstocks development, (B) Biofuels and biobased products development, and (C) Biofuels development analysis. The intent of requiring integration of the three areas is to encourage a collaborative problem-solving approach to all studies funded under BRDI, to facilitate formation of consortia, identify and address knowledge gaps, and accelerate the application of science and engineering for the production of sustainable biofuels, bioenergy and biobased products.

All of these USDA programs, including research, strategically help to address the immediate and short-term needs for the production of the feedstocks necessary to meet the next 21 billion gallons of biofuels under the Renewable Fuel Standard (RFS2), including the needs for drop-in aviation and military fuels.

RESPONSE TO WRITTEN QUESTIONS SUBMITTED BY HON. MARIA CANTWELL TO RICHARD ALTMAN

Question 1. Mr. Altman, do you believe that the largest remaining obstacle for the creation of a functioning market in aviation biofuels is the ability of key participants in the supply chain to secure the financing necessary to develop commercially useful volumes of alternative fuel? A number of witnesses spoke about their expectation that ATSM will certify alcohol-to-jet fuel specification in the next few years. Do expectations regarding the future alcohol-to-jet fuel pathway impact the ability for key participants in HRJ fuel pathway supply chains to secure financing today?

Answer. Yes, securing financing is the key to success going forward. Key financing agencies backed by the U.S. governments (such as SBA, and trade groups) must view qualified aviation fuels technologies as "proven." Right now funding is constrained because the government agencies backing loans will not fund biofuel facilities.

No HRJ facilities are not jeopardized by the expectations of ATJ pathways being developed nor has it been an issue with IPO's for HRJ companies to my knowledge. Assuring affordable feedstocks by increasing yields for HRJ (HEFA) feedstocks is the key to meeting HRJ potential. ATJ qualification is needed to improve supplies to allow us to eliminate imported oil dependence.

Question 2. Mr. Altman, one of CAAFI's most significant accomplishments was developing the Fuel Readiness Level for different pathways and a Feedstock Readiness Tool. Sustainable Aviation Fuels Northwest Identified four potential feedstocks for the region—oilseed crops such as camelina, forest residue, municipal and industrial solid wastes, and algae production. Where do they rank relative to each other in terms of readiness?

Answer. Oilseed crops such as camelina are ready now. Forest residue waste for processing via fischer tropsh processes are ready now for commercial use.

Municipal waste to the degree it requires pre treatment of the waste to separate out unusable parts or to make them useable is more risky from a production readiness perspective. I trust that projects such as the Solena, British Airways project will handle that those issues at FRL levels beyond certification.

Algae is actually an oil seed crop so FRL is not an issue. What is an issue is the economics and energy intensity of open and closed pond systems to produce both economic and environmentally acceptable results associated with such issues as water extraction. This too is a production readiness challenge. I personally have not seen the evidence that we are there in these areas. Sufficient research funding is in place in my view through DARPA and DOE to address this challenge but it is not clear when or if the challenge will be met. I do not think that we should depend on algae economics coming home to address the biofuels challenge with any certainty. Some technologies (*e.g.,* hydrogen fusion) simply do not come home in a timely manner to address concerns. I am hopeful that this is not the case with Algae . . . on a scaled producible basis. But there is a risk that this is the case according to the experts that I have spoken to.

Question 3. Mr. Altman, AltAir Fuels was able to get over a dozen airlines to agree to accept up to 750 million gallons of alternative fuel over 10 years. In 2009, airlines operating at SeaTac consumed 411 million gallons of jet fuel.

Answer. Not sure what the question is here but the statement is accurate to the degree that the intent was established via MOU. Definitive terms have yet been signed but we are helpful and supportive of the Altair project as a "first of its kind" production facility in the U.S.

Question 4. There are different business models how fuel is stored and distributed at airports. At some airports, the tenant airlines form a fueling consortium and then hire a contractor to manage it. At other airports, individual airlines are responsible for managing their own fuel needs. Then there are those airports where the airport runs everything. Do you believe the prevailing business model at an airport for storing and distributing aviation fuel will impact the adoption of aviation biofuels by commercial airlines operating at the airport?

Answer. I believe that this business model is helpful in that allows all the airlines at the airports to share the risk of new product introduction. It also assures the broadest acceptance across the industry.

Question 5. Mr. Altman, the military tries to simplify some of its logistics by having some of its ground support vehicles at its bases operate on jet fuel. The EPA won't let Jet A be used for ground service equipment at commercial airports because of the amount of sulfur in the fuel. Aviation biofuel does not contain sulfur. If the EPA would allow green jet fuel to be used to run ground service equipment, would that help increase demand by airports?

Answer. This is more a question for ATA fuel buyers than for me. I encourage you to ask Ms. Pinkerton.

There are more factors than EPA however.

For example we need to look at the taxation rules for GSE fuel and jet fuel I understand that they are different and may limit the use of Jet fuel.

I also think it important to the economics of airport use of fuel that the airports or airlines be permitted to distribute fuel diesel fuel use at the airport off-airport without consequence for projects funded by airport programs. My understand that this is a concern if we are to maximize the ability to use the airports as a concentrated distribution hub. I hope that we will investigate and expose this issue more in case studies that we will be performing in the ACRP 02–36 project under the Airport Cooperative Research program.

Question 6. Mr. Altman, one of the reasons the approved HRJ fuel specification is a 50–50 blend rather than 100 percent biofuel is that the chemical composition doesn't react with the rubber seals in a way to ensure there is no fuel leakage. One alternative discussed is using seals made out of a different material. How complicated and expensive would it be to identify, locate, and replace seals throughout the fueling infrastructure and on aircraft? What is the potential of developing a pathway that would allow for a 100 percent biofuel rather than a blend?

Answer. To date the need to qualify 100 percent biofuels has been academic in nature in that there has not been adequate supply to support facilities at the 100 percent level. Hence that portion of the agenda has been put on hold.

Presently if fuel with less than 8 percent aromatic content (risk occurs with >50% HRJ or FT is used) is an environmental hazard associated with fuel leaks. The issue of whether it is more economic to change out seals or to require additives rich in aromatics (*e.g.,* pyrolysis oil derive additives) has not been evaluated to my knowledge.

In my view the Committee could offer to fund such an evaluation in conjunction with the increased support for advanced biofuels research similar to what was put in place in FY 10 for the advance biofuels supplemental funds which you approved. It would not be very expensive to do that evaluation in my view.

RESPONSE TO WRITTEN QUESTION SUBMITTED BY HON. JOHN D. ROCKEFELLER IV TO JOHN PLAZA

Question. Can you discuss broadly how the development of alternative fuels creates job growth across the supply chain?

Answer. Imperium Renewables was founded in 2004 and operates a 100,000,000 gallon per year biofuel facility.

i. Since 2004 our company has contributed the following to the economy:

1. We have employed over 150 people
2. Provided over $12,000,000 in payroll benefits to our employees in Washington State
3. Provide jobs and opportunity for veterans returning home from Iraq and Afghanistan at our biofuel facility
4. Invested over $135,000,000 into Washington state companies for construction and operation of our biofuel facility
5. Purchased over $350,000,000 of agricultural products from the U.S. and Canada
6. Sold over 85,000,000 gallons of biodiesel for more than $340,000,000
7. Offset over 1.35 billion lbs of CO_2 (per National Biodiesel Board)
8. In total, our company has contributed over $940,000,000 to the North American economy since 2004

ii. With the enactment of EISA 2007 RFS2 and the creation of a 36 billion gallon per year market by 2023, our company is just 1/360th of the potential economic benefit to the U.S.

iii. By adding the Department of Defense as a market for biofuel producers, this economic engine can be accelerated and enhanced to bring significantly more jobs than already are in place with this burgeoning industry

iv. Imperium's supply chain includes oilseed growers in rural parts of the country, crushing facilities that extract the oils from the feedstock, as well as transportation of raw materials to the facility and of finished products from the facility. As we expand our production, our demand for goods and services throughout the supply chain will naturally increase. According to a 2009 industry report by Bio Economic Research Associates, "direct job creation from U.S. advanced biofuels production could reach 29,000 jobs by 2012, rising to 94,000 by 2016, and 190,000 by 2022." In addition, the report found that "total job creation, accounting for economic multiplier effects, could reach 123,000 jobs in 2012, 383,000 in 2016, and 807,000 by 2022."

Center for a New American Security—September 2010

FUELING THE FUTURE FORCE

PREPARING THE DEPARTMENT OF DEFENSE FOR A POST-PETROLEUM ERA

By Christine Parthemore and John Nagl

Acknowledgments

We would like to thank our colleagues at the Center for a New American Security (CNAS) for their valuable insights and comments throughout the research and writing process. Will Rogers, Dr. Kristin Lord, and more than a dozen colleagues all provided invaluable feedback and critiques. Joseph S. Nye, Jr. National Security Intern Alexandra Stark contributed her sharp, investigative research skills and excellent writing. We are grateful for external reviews of drafts from, among others, CDR Herb Carmen, USN, Frank Hoffman of the Navy Staff and Jim Morin of Hogan Lovells. As always, Liz Fontaine, Ashley Hoffman and Shannon O'Reilly provided guidance and advice through the production process. Many experts from the U.S. Navy, Air Force, Army, Marine Corps, and the Office of the Secretary of Defense, and other U.S. Government agencies and NGO's, contributed to the discussions from which we derived this analysis; however we alone are responsible for any errors or omissions.

I. Introduction

The U.S. Department of Defense (DOD) must prepare now to transition smoothly to a future in which it does not depend on petroleum. This is no small task: up to 77 percent of DOD's massive energy needs—and most of the aircraft, ground vehicles, ships and weapons systems that DOD is purchasing today—depend on petroleum for fuel.[1] Yet, while many of today's weapons and transportation systems are unlikely to change dramatically or be replaced for decades, the petroleum needed to operate DOD assets may not remain affordable, or even reliably available, for the lifespans of these systems.

To ready America's armed forces for tomorrow's challenges, DOD should ensure that it can operate all of its systems on non-petroleum fuels by 2040. This 30-year time-frame reflects market indicators pointing toward both higher demand for petroleum and increasing international competition to acquire it. Moreover, the geology and economics of producing petroleum will ensure that the market grows tight long before petroleum reserves are depleted. Some estimates indicate that the current global reserve-to-production (R/P) ratio—how fast the world will produce all currently known recoverable petroleum reserves at the current rate of production—is less than 50 years.[2] Thus, given projected supply and demand, we cannot assume that oil will remain affordable or that supplies will be available to the United States reliably three decades hence. Ensuring that DOD can operate on non-petroleum fuels 30 years from today is a conservative hedge against prevailing economic, political and environmental trends, conditions and constraints.

It will take decades to complete this transition away from petroleum. However, DOD has already laid important groundwork. The development, testing and evaluation of renewable fuel conducted by the armed services to date mark the first steps in guaranteeing DOD's long-term ability to meet its energy needs. DOD should build on this work and develop a strategy that guarantees its ability to operate worldwide in the event of petroleum scarcity or unavailability.

The Center for a New American Security (CNAS) launched a project in September 2009 to examine DOD's energy challenges and recommend a path forward. We convened DOD leaders and nongovernmental experts; researched current laws, requirements and projects; and visited military bases around the country to discuss DOD's energy challenges and opportunities. From this research, we concluded that DOD needs a long-term strategy to adopt alternative fuels based on our reading of current trends in petroleum availability and use, as well as our identification of petroleum dependence as a long-term vulnerability for DOD.

DOD officials increasingly understand this vulnerability. During the course of our project, the Navy appointed two-star officers to lead two task forces on energy and climate change. Their activities, which began quietly within the bureaucracy, are now well-known examples of leadership by the U.S. armed forces. The Air Force and Navy flight-tested camelina-based biofuel blends in the past year.[3] The Air Force's Air Mobility Command and the Office of the Secretary of Defense (OSD) are working to increase energy efficiency and maximize fuel savings in existing platforms and new acquisitions. The Quadrennial Defense Review (QDR) presented instructions for integrating energy considerations into how DOD does business. Bases around the country are investing in solar, wind and geothermal projects. DOD is working to comply with federal energy mandates, and in particular those found in the Energy Independence and Security Act (EISA) of 2007, President Barack Obama's October 2009 Executive Order on resource conservation by federal agencies and defense authorization acts.

Though each of the services has admirably developed its own energy strategy to improve its near-term energy management, DOD must also develop a comprehensive long-term energy strategy. The strategies developed by individual services focus heavily on electricity usage at domestic installations, which accounts for a relatively small fraction of DOD's energy needs, and most goals within these strategies do not look beyond 2015 or 2020—a timeline that is too short to ensure DOD's long-term energy security. Moreover, there is no single official who oversees DOD's entire energy portfolio; authority within DOD is currently divided, which is likely to complicate implementation of the strategy. This report lays out the strategic necessity

---

[1] U.S. Energy Information Administration, *Annual Energy Review 2009*, "Table 1.13: U.S. Government Energy Consumption by Agency and Source, Fiscal Years 2003, 2008 and 2009." (19 August 2010); and BP, *Statistical Review of World Energy* (2010).

[2] Energy Information Administration, "Crude Oil and Total Petroleum Imports Top 15 Countries" (May 2010 Import Highlights); and BP, *Statistical Review of World Energy* (June 2010).

[3] Jason Paur, "Air Force Debuts Biofuel-Guzzling Warthog," *Danger Room* (30 March 2010); and Liz Wright, "Navy Tests Biofuel-Powered 'Green Hornet'," Official Website of the United States Navy (22 April 2010).

for DOD to find alternatives to petroleum over the next 30 years and then presents important steps in achieving that long-term goal.

Transitioning away from petroleum dependence by 2040 will be enormously difficult, but fortunately the U.S. defense sector has made several energy transitions successfully in its history. In particular, it moved from coal to petroleum to nuclear power in its ships. In a similarly seismic shift, DOD rapidly increased its reliance on electronics, space assets and computer systems in modern warfare in ways that enhanced mission effectiveness. These experiences may offer lessons for DOD as it leverages an energy transition to maximize its strategic flexibility and freedom of maneuver.

Now is an opportune time to make this transition. As the services redeploy from current wars, the Army (and to a lesser extent the other services) have years of reset ahead of them. Acquisition reforms and personnel restructuring initiatives launched by Secretary Robert Gates in 2009 and 2010 will continue through the Obama administration and likely beyond. Together, these developments will present opportunities to procure new, more energy-efficient systems.

A successful transition away from petroleum will produce financial, operational and strategic gains. Reducing dependence on petroleum will help ensure the long-term ability of the military to carry out its assigned missions—and help ensure the security of the Nation. Though adopting nonpetroleum fuels will require an initial investment, it will likely be recouped in budget savings over the long term. Finally, moving beyond petroleum will allow DOD to lead in the development of innovative technologies that can benefit the nation more broadly, while signaling to the world that the United States has as innovative and adaptable force.

This transition should not compromise readiness and, indeed, DOD must always put mission first. However, DOD need not choose between accomplishing its mission and minimizing the strategic risks, price fluctuations and negative environmental effects of petroleum consumption. By providing the private sector with stable market signals and incentives to invest in scaling up the fuels that meet its unique energy needs, DOD will never need to sacrifice performance or national security for energy security. Rather, reducing reliance on petroleum will only help the armed services to accomplish their missions in the years and decades to come.

Table 1: DOD Energy Consumption by Fuel, 2009

2009 DOD Energy Consumption by Fuel Source, in Trillion British Thermal Units (BTU)		
Fuel Source	Energy Use	Percentage of total
Petroleum	679.7	77.2
Natural Gas	74.2	8.4
Coal	16.2	1.8
Chilled water, renewable energy, and other fuels reported as used in facilities	9.1	1.0
Other electric	101.1	11.4
Total	*880.3*	*99.8*

Source: Department of Energy, "U.S. Government Energy Consumption by Agency and Source, Fiscal Years 2003, 2008 and 2009." Totals may not equal 100 percent due to rounding.

II. Why DOD Should Adopt Alternative Fuels

Several factors challenge DOD's continued reliance on its existing petroleum-dominant energy strategy over the long term: direct risks to U.S. security; troubling supply and demand trends; the often-hidden external costs of fuel consumption; and a changing domestic political and regulatory environment.

The Risks of Petroleum Dependence

The growing world demand for petroleum presents major geostrategic risks. High prices and rising demand are a boon to major suppliers and reserve holders such as Iran and Venezuela, which are unfriendly to the United States. It also affects the international behavior of rising powers such as China, which is on a quest to secure access to natural resources that is in turn expanding its influence around the globe. In Mexico, one of the top suppliers of petroleum to the United States, pipelines serve as an increasingly attractive target for dangerous cartels to fund activities that could undermine the Mexican government, destabilize the region and decrease U.S. homeland security.[4] American foreign policy itself has been colored by its growing petroleum demands since the 1970s oil crises and subsequent declaration of the Carter doctrine, which stipulated that the United States would consider threats to the Persian Gulf region threats to its "vital interests" due to the strategic importance of its petroleum reserves.[5]

Dependence on petroleum for 94 percent of transportation fuel is also a dangerous strategic risk for the United States given the leverage oil can provide to supplier countries. Many European allies have experienced such leverage in action with Russia periodically threatening to reduce or cutoff natural gas exports to countries highly reliant on their supplies (and in some cases carrying through with these threats). Similarly, national oil companies and OPEC can choose to increase or decrease their production rates to drive changes in the market.

The more the United States reduces its dependence on petroleum, the better it can hedge against petroleum suppliers exerting political leverage over U.S. interests, including in times of crisis.

At the operational level, heavy reliance on liquid fuels also constitutes a force protection challenge for DOD. Fuel supply convoys have been vulnerable to attack in both Iraq and Afghanistan, where the services have struggled to adapt to the challenges of terrorism, insurgency and violent extremism. In addition to minimizing these risks in the current wars, DOD must also conceptualize and plan for what the future will likely hold for America's security. The Navy's battle against pirates off the coast of the Horn of Africa foreshadows the littoral and unconventional challenges that await the United States in the coming decades, as populations continue to migrate toward the world's coastal area. These types of problems often manifest at major shipping chokepoints (including petroleum transit chokepoints), and addressing them will include distinctive fueling requirements. The Air Force, likewise, confronts dramatic changes in manned and unmanned flight, in addition to the proliferation of space technologies, all of which could dramatically alter fuel needs. In another example, one recently published AirSea battle concept focused on China notes that the type of conflict it outlines could require hardening fueling infrastructure, improving aerial refueling, "stockpiling petrol, oil, and lubricants" and potentially "running undersea fuel pipelines between Guam, Tinian and Saipan."[6] As the character of warfare changes, DOD will have to continue to consider the attraction of fuel supply lines to opponents.

Changing Supply and Demand

DOD cannot be assured of continued access to the energy it needs at costs it can afford to pay over the long term. Today DOD meets its energy needs primarily through petroleum, which accounts for more than 77 percent of DOD's total energy use.[7] However, both demand and supply trends are likely to raise the price and perhaps even limit the availability of petroleum.

The U.S. Energy Information Administration projects that world energy demand will grow from its 2007 level of 495.2 quadrillion British thermal units (Btu) to 738.7 quadrillion Btu by 2035—a steep increase. If current trends continue, energy demand in non-OECD countries will grow more than four times faster than in OECD countries.[8] Global petroleum demand has increased steadily from about 63 million barrels of oil per day in 1980 to more than 85 million barrels today, and will grow to 110.6 million barrels per day by 2035 if current trends hold.[9]

While global oil demand increases, the supply side of the equation is equally worrisome. At current production rates, the global R/P ratio is about 46 years (see *Ap-*

[4] Steve Fainaru and William Booth, "Mexico's Drug Cartels Siphon Liquid Gold: Bold Theft of $1 billion in Oil, Resold in U.S., Has Dealt a Major Blow to the Treasury," *The Washington Post* (13 December 2009).

[5] President Jimmy Carter, "State of the Union Address" (23 January 1980).

[6] Jan van Tol, "AirSea Battle: A Point-of-Departure Operational Concept" (Washington: Center for Strategic and Budgetary Assessments, 2010): 81–82.

[7] U.S. Energy Information Administration, *Annual Energy Review 2008*.

[8] U.S. Energy Information Administration, *Annual Energy Outlook 2010*, Table A1: "World total primary energy consumption by region, Reference case, 2005–2035" (11 May 2010).

[9] U.S. Energy Information Administration Website, "Petroleum Statistics" (26 January 2010); *Annual Energy Outlook 2010*, Appendix A.

pendix I). Proved reserves (those recoverable under current conditions[10]) increasingly lie in the hands of national oil companies that are often hostile to U.S. interests. Venezuela, for example, holds over 100 years' equivalent of reserves at its current production rates. Thus, the U.S. reliance on countries such as Venezuela as a supplier could increase beyond the roughly 1 million barrels of petroleum it already imports from there every day.[11] The reserve part of this ratio may increase, but we can also be certain that the demand half of the ratio will increase, and likely at a faster pace.

Costs of Petroleum Dependence

- Heavy dependence on large fuel supplies can increase operational vulnerabilities and make fuel supply infrastructure a more valuable target.
- Every dollar increase in the price of petroleum costs DOD up to 130 million additional dollars.
- Rising global demand, for instance in China, is increasing the strategic importance of petroleum in ways that could be detrimental to U.S. interests.
- Countries such as Iran and Venezuela could have the largest remaining reserves in a few decades if current production rates hold—and will gain leverage as a result.
- High levels of petroleum consumption are contributing to the changing climate, which can bring destabilizing effects and trigger new security challenges.

The United States is already moving past the era of nearly complete reliance on petroleum for transportation fuel. Though it will take several decades to make this transition, the country should take every opportunity to hasten progress given projections of tight markets and a heightened potential for competition. This transition will require careful investments that account for the potential economic, environmental and geopolitical tradeoffs involved with all energy sources.

There is an array of reliable, renewable fuels that should be considered as alternative supplies to petroleum, including multiple generations of biofuels. Biotechnicians have long proven the technical ability to produce hydrocarbon equivalents to fossil fuels, including the jet fuel blends that DOD requires. Efforts by the National Laboratories, academia and the private sector are focusing on basic science that will enable more efficient use of second- generation biological fuel sources (made from non-food crops) by increasing efficiency in processing plant materials while retaining net energy gains, and by overcoming other technical hurdles. Others are leapfrogging beyond second-generation biofuels to fuels derived from algae. Still other options include displacing petroleum by using electricity or natural gas to power transportation, and using distributed renewable energy at overseas and forward operating bases to displace petroleum in powering generators. It is encouraging that growth in renewable energy supply availability frequently outpaces expectations. Ethanol production grew 164 percent between 2002 and 2006, and biodiesel production expanded from 1 trillion Btu to 32 trillion Btu over the same period. Wind, solar and geothermal supplies also have expanded faster than most analysts predicted over the past decade.[12] These supply-side changes show how technical, economic and policy decisions, such as tax regimes that Congress has enacted to even the playing field with fossil fuels, can affect energy trends.

Any effective DOD energy strategy must also be flexible enough to account for the fact that its leaders will have to make energy decisions based on imperfect information. Specific projections regarding how rapidly fuel alternatives could achieve large-scale production and consumption are often treated as proprietary. This uncertainty is particularly problematic for DOD, which has limited manpower and funds to invest in fuel research and development.

The Indirect Costs of Petroleum Dependence

The Department of Defense accounts for about 80 percent of the Federal Government's energy consumption, and its high dependence on petroleum-based fuels—the Defense Energy Support Center reported 132.5 million barrels in petroleum sales

[10] The U.S. Department of Energy defines "proved reserves" as follows: "Proved reserves are estimated quantities that analysis of geologic and engineering data demonstrates with reasonable certainty are recoverable under existing economic and operating conditions." U.S. Energy Information Administration, "World Proved Reserves of Oil and Natural Gas, Most Recent Estimates" (March 2009).

[11] Energy Information Administration, "Crude Oil and Total Petroleum Imports Top 15 Countries" (May 2010 Import Highlights); BP, *Statistical Review of World Energy* (June 2010).

[12] U.S. Energy Information Administration Website, "Biomass" (April 2008).

in Fiscal Year 2008, totaling nearly $18 billion [13]—means that its budget is subject to major oil price fluctuations.[14] Petroleum price spikes negatively affect DOD's budget and divert funds that could be used for more important purposes. As Secretary Gates said in 2008, "Every time the price of oil goes up by one dollar per barrel, it costs us about $130 million." [15] In an era of constrained budgets, American security is best served by trying to hedge against future price fluctuations of this scale.

In addition to the security and financial costs, petroleum dependence creates environmental costs that are causing increasing concern among security analysts. Emissions from fossil fuel use contribute to changes in the global climate, which risk altering geopolitical relations, destabilizing regions of high strategic importance to the United States, increasing erosion and storm surges at coastal installations, and altering disease patterns.[16] Melting summer ice in the Arctic is an early example; its geopolitical importance has risen sharply in the past 5 years as Arctic countries (and their potential shipping and natural resource customers) prepare to exploit newly navigable waterways and seabed resource deposits. federal leaders from both major political parties, DOD's civilian and military leaders, and security analysts of all stripes regularly reiterate concerns over the national security implications of the changing climate caused by high-carbon fuel consumption.[17] Other environmental costs of fuel production can include heavy water use and diverting arable land to fuel production, both of which can trigger negative side effects if not managed properly. Factors such as greenhouse gas emissions (including from burning high-carbon fuels and from land use change) and the effects of fuel production on food prices should therefore constrain DOD's energy investments in high-carbon fossil fuels or first-generation biofuels derived from food crops.

The Changing Political, Legal and Regulatory Environment

Signs indicate that federal and state governments will continue to push for greater adoption of domestic and/or lower-carbon energy technologies. As a result, DOD will face a changing legal, regulatory and political environment in the coming decades. Congress has consistently passed legislation since 2005 to support investments and set federal requirements supporting energy efficiency and renewable energy production. The Obama administration strongly supports this approach as well. Obama issued an October 2009 Executive Order committing federal agencies to calculate and reduce their greenhouse gas emissions, which spurred energy-focused DOD officials to begin complying with this requirement. Likewise, 27 states have instituted renewable energy portfolio standards, and nine others have renewable or alternative energy goals or requirements.[18] Legal and regulatory changes can also constrain energy choices. For instance, the U.S. Supreme Court ruled in 2007 that greenhouse gas emissions constitute a pollutant and therefore can be regulated at the federal level, and the Obama administration has signaled its intent to move forward with such regulation unless the Congress mandates emissions reductions through legislation.

While the U.S. Government sets domestic regulations and laws, and can exempt combat-related activities, it does not exercise the same control internationally. In-

---

[13] Defense Energy Support Center, *Fact Book FY 08* (2009): 19–20.

[14] U.S. Energy Information Administration, *Annual Energy Review 2009*, "Table 1.13: U.S. Government Energy Consumption by Agency and Source, Fiscal Years 2003, 2008 and 2009." (19 August 2010).

[15] Donna Miles, "Military Looks to Synthetics, Conservation to Cut Fuel Bills," *American Forces Press Service* (6 June 2008).

[16] See, for example, Commander Herbert Carmen, USN, Christine Parthemore and Will Rogers, *Broadening Horizons: Climate Change and the U.S. Armed Forces* (Washington: Center for a New American Security, 2010).

[17] Many U.S. policymakers, agencies and documents have recognized the connection between climate change and security: the 2010 QDR says "Climate change and energy will play significant roles in the future security environment": p. xv; the National Intelligence Council has done extensive climate change research including reports on "The Impact of Climate Change to 2030"; President Obama stated in his December 2009 Nobel Peace Prize acceptance speech that ". . . the world must come together to confront climate change. There is little scientific dispute that if we do nothing, we will face more drought, famine and mass displacement that will fuel more conflict for decades. For this reason, it is not merely scientists and activists who call for swift and forceful action—it is military leaders in my country and others who understand that our common security hangs in the balance"; in remarks in November 2009 Secretary of Defense Robert Gates said that "the melting of the polar ice cap in the Arctic plus the frequency and intensity of weather events in this hemisphere, with the corresponding need for military humanitarian assistance missions, calls for a greater attention to the security implications of climate change."

[18] Pew Center on Global Climate Change, "Renewable & Alternative Energy Portfolio Standards" (14 December 2009).

deed, there is growing concern that foreign countries may not always exempt military activities within their territory from environmental standards. For example, the Canadian government recently decided to upgrade one of its vessels that was not equipped to meet the environmental standards of several European countries, for fear that the vessel could be denied port access.[19] The Department of Defense must consider emerging international trends in regulating emissions and adopting less carbon-intensive energy sources as it considers how to guarantee its freedom of access to foreign ports and territories.

III. Elements of a DOD Energy Strategy

In response to these factors, DOD should map a path forward that relies on technological innovation and efficiency to hedge against price spikes and scarcities and to accommodate America's economic, political and environmental needs. By planning now around these likely future conditions, DOD can weather change, protect its own interests, reduce its vulnerability to extreme price spikes and—most importantly—ensure that it can meet its mandate to protect the Nation's security. The logical next step is to develop a strategy that adheres to 12 specific guiding principles.

1. Set a Common Energy Goal

In order to address security risks, costs, domestic constraints and changing energy supply and demand trends, DOD should set an overarching energy goal of *managing a smooth transition beyond petroleum over the next 30 years*. This goal is significantly broader than the array of goals and objectives that the services have set to guide their own energy decisions to date. Those more near-term goals move in the right direction, but remain insufficient given the broad scope and extended timeline of DOD's energy challenges.

The 2010 QDR stated, "Energy security for the Department means having assured access to reliable supplies of energy and the ability to protect and deliver sufficient energy to meet operational needs."[20] This leaves much room for interpretation and is not precise enough to ensure that everyone within DOD is moving in the same direction. To many domestic installations, energy security means reliable sources of power that are not vulnerable to disruption by natural or man-made disruptions affecting the electric grid. To the Army, operational needs and installation energy concerns overlap greatly given that operations abroad center most often on forward operating bases. The Air Force is yet a different case; as aviation fuel accounts for the majority of its energy demand, liquid fuel supplies are of paramount importance. Thus, for each of the services, the broad requirements of "assured access," "reliable" and "sufficient" supplies could mean any number of energy choices, and will vary depending on whether this definition applies to short-term or long-term needs.

To accommodate all of these needs, yet still provide real guidance, DOD should settle on a single overall goal and ensure that the objectives set by the services align with that goal. It is important that this goal is long-term in nature and general enough to incorporate the work already set by the military services and to allow flexibility, but specific enough to guide real changes in behavior and investment.

A Thirty-Year Challenge

We recommend that DOD establish a goal that by 2040, DOD must be able to operate all of its assets on non-petroleum fuels. The thirty-year timeline is sufficient time for the private sector scaling up adequate supplies, and for DOD aligning its bureaucratic and infrastructure systems to accommodate this change. Knowing that petroleum prices will rise and renewable fuels will become cost-competitive years before the world produces all reserves, it is not prudent to assume that petroleum will remain affordable or that supplies will be reliably available to the United States three decades hence; nor is it wise to perpetuate the geopolitical, operational and environmental costs indefinitely. Ensuring that DOD can operate on non-petroleum fuels 30 years from today is therefore a conservative hedge against the economic, political and environmental conditions and constraints outlined in this report.

Despite the 30-year timeline, DOD does not have several decades to *begin* this transition. The renewable fuel development, testing and evaluation that the services have conducted to date mark the first steps in guaranteeing their long-term ability to meet their energy needs, but even if DOD adopts a hastened timeline, it will take decades to complete this transition. Implementing this strategy must therefore begin immediately.

---

[19] Bill Curry, "Canadian Navy's Ships Risk Being Banned from Foreign Ports," *The Globe and Mail* (Toronto) (5 August 2010).

[20] QDR: 87.

Though it is important to start the critical process of transitioning to non-petroleum energy sources, mission accomplishment will always remain DOD's top consideration. It is therefore essential that DOD's energy choices do not interrupt or detriment operational capabilities. Rear Admiral Philip Hart Cullom, director of fleet readiness for the Navy staff and head of the Navy's Task Force Energy, calls this creating "off-ramps" from petroleum.[21] In the near term, this indicates the importance of drop-in fuels, or liquid fuels that are chemically equivalent to petroleum-based fuels and can therefore fuel existing platforms. DOD's energy transition should be nearly seamless to the soldiers, sailors, airmen and Marines using these fuels.

Other goals debated in recent years, including a goal of simply increasing the efficiency of petroleum use or a static reduction in overall fuel consumption, will be insufficient. Improving energy efficiency—in other words, getting more power per unit of energy consumed—must be part of a strategy to meet DOD's energy needs without petroleum, but it is important that this not serve as the goal itself. Efficiency is one of the most important short-term operational energy objectives for DOD; for instance, any energy efficiency gains in Iraq and Afghanistan can immediately reduce vulnerable supply lines, save lives and free up manpower for other operations. However, efficiency does not mark a concrete end state over a multidecade time scale, and therefore cannot serve as an overarching goal. America's energy efficiency has grown since the 1970s, yet its overall petroleum demand and corresponding vulnerabilities have also grown. For DOD, this means that its operational vulnerabilities and costs remain despite its efficiency gains. In other words, gains in efficiency are necessary and important, but there is a danger that too heavy a focus on efficiency over a long-term time scale will mask an increasing reliance on fuel that poses further risks to the Department of Defense. Efficiency should therefore be treated as a means and an operational enabler.

Service Priorities

The services have set many of the necessary short- and near-term goals and objectives to hit our suggested long-term target for DOD as a whole. The Army, Air Force, Navy and Marine Corps all established energy strategies, and they have since refined them to accommodate new requirements from Congress and executive orders. These include, among others:

Air Force

"By 2016, be prepared to cost competitively acquire 50 percent of the Air Force's domestic aviation fuel requirements via an alternative fuel blend in which the alternative component is derived from domestic sources produced in a manner that is greener than fuels produced from conventional petroleum."[22]

"Test and certify all aircraft and systems against 50/50 alternative fuel blend by 2011."[23]

"Reduce overall fossil fuel consumption in vehicles by 2 percent annually (2005 baseline) until 2015, and steadily increase the overall fleet average miles per gallon (MPG)."[24]

"Install at least 1 renewable fuel pump at each federal fleet refueling center at each installation that issues more than 100 thousand gallons of ground fuel annually."[25]

Army

"Reduce the amounts of power and fuel consumed by the Army at home and in theatre. This goal will assist in minimizing the logistical fuel tail in tactical situations by improving fuel inventory management and focusing installations consumption on critical functions."[26]

"Raise the share of renewable/alternative resources for power and fuel use, which can provide a decreased dependence upon conventional fuel sources."[27]

[21] Rita Boland, "Great Green Fleet Prepares to Set Sail," *SIGNAL Magazine* (July 2010).
[22] Air Force Energy Plan (2010): 8.
[23] Ibid.
[24] Air Force Infrastructure Energy Plan (2010): 12.
[25] Ibid.: 14.
[26] Army Senior Energy Council and the Office of the Deputy Assistant Secretary of the Army for Energy and Partnerships, "Army Energy Security Strategy" (13 January 2009): 4.
[27] Ibid.

Navy

"The Navy will demonstrate in local operations by 2012 a Green Strike Group composed of nuclear vessels and ships powered by biofuel. And by 2016, we will sail that Strike Group as a Great Green Fleet composed of nuclear ships, surface combatants equipped with hybrid electric alternative power systems running biofuel, and aircraft flying only biofuels—and we will deploy it."[28]

"The Department of the Navy will by 2015 reduce petroleum use in our 50,000 strong commercial fleet in half."[29]

Marine Corps

"Reduce energy intensity 30 percent by 2015 relative to a 2003 baseline."[30]

"Increase the percentage of renewable electrical energy consumed to 25 percent by FY 2025."[31]

Once DOD establishes its long-term energy goal, it will need to audit these energy plans to ensure that all service-level energy goals align. Most, if not all, of them will already align with the long-term goal of managing a smooth transition beyond petroleum by 2040. It will be critical to build on these successes by expanding targets past the dates specified above.

It is also important that DOD's energy goal does not amount solely to absolute reductions in energy consumption, devoid of consideration of how DOD uses energy in its efforts to protect and defend U.S. interests. DOD must always retain the flexibility to successfully conduct its missions. Demand reduction can be an important means of reducing vulnerabilities to supply lines abroad and reliance on a fragile grid at home. However, overall energy consumption should remain a function of DOD's activities and global engagements. Total fuel demand must therefore remain flexible and should not serve as a fixed, long-term goal.

2. Establish Clear Energy Guidelines for DOD

DOD should establish, publish and enforce a clear set of overarching rules or guidelines to help the services navigate their energy transitions, and to signal to the private sector what sorts of fuels, infrastructure and efficiency technologies it will need to supply over the long term.

In setting these guidelines, first and foremost, DOD's *energy investments must meet military needs*. Those that cannot be designed or adapted by their producers to meet military needs should not be considered worth DOD's limited energy investment dollars. Otherwise, as the track record to date indicates, new fueling infrastructure, energy production technologies and vehicles will simply not be used. For example, a hydrogen vehicle and fueling station demonstration at Hickam Air Force Base in Hawaii marked a great sign that DOD bases can be used for testing new technologies, but the small scope of the demonstration—a single fueling station and limited range of the vehicles—significantly limited the utility of this investment to the airmen and civilians working at Hickam. DOD's purchases should treat military utility as a mandatory constraint on any energy-related purchases.

Second, the fuels on which DOD relies *must be consistently available long into the future*. This stipulation leads to a preference for renewable fuel technologies versus supplies that will eventually deplete. We do not currently know with much fidelity what energy supplies will be reliably available where and when—even for petroleum beyond the 30-year timeframe, with the likelihood of demand spiking, possible recalcitrance on behalf of suppliers, diminishment of easily recoverable supplies and fragile transit routes and delivery infrastructure. DOD requires consistently available supplies and supply systems that will not evaporate for economic or political reasons.

Third, *new fuel sources must hold the potential to be available globally*. DOD relies on international companies and other countries to provide fuel supplies for its use outside of the United States. Reliance on a single fuel that is commonly used in all countries and produced globally (petroleum) benefits DOD logistically, but this system will not survive indefinitely at a bearable cost. Many countries are already producing fuel alternatives to petroleum and increasing their capacity to do so, though there is a lack of information about where these supplies are, whether they

[28] Secretary of the Navy Ray Mabus, "Remarks at the Naval Energy Forum" (14 October 2009): 8–9.

[29] Ibid.

[30] United States Marine Corps, "Ten by '10: Top 10 Things To Do by 2010 to Reduce USMC Energy Risks" (2009): 3.

[31] Ibid.

can be formulated to fit DOD's technical specifications, and to what scale they are likely to grow in supply availability. DOD must insist that its platforms can operate on fuels that it can procure abroad in order to ensure its ability to operate globally and to take advantage of the benefits that fuel source diversification can offer.

Fourth, *performance is paramount.* DOD cannot waver on its demand for fuels that perform properly. Its assets, particularly aircraft, require chemical consistency in the fuels used. This indicates special concern for reliability in formulating, refining and properly blending drop-in aviation biofuels that are mixed with petroleum.

Fifth, plans to smoothly navigate DOD's long-term energy transition beyond petroleum *must account for the changing political and regulatory environment.* Congress mandated that federal agencies should not invest in fuel sources that carry lifecycle greenhouse gas emissions higher than their current fuel sources[32] (such as coal-derived fuels, and by some calculations, potentially corn-based ethanol), and private companies are pushing Congress more aggressively than ever to enact legislation to curb emissions. No federal agencies should invest their finite funds in fuel sources with higher lifecycle emissions, or that contribute to extensive damage to food commodity markets or ecosystems. The private sector should not sell DOD fuels that will contribute to extensive rainforest destruction, water supply contamination or climate-changing emissions increases. DOD needs to set and stick to guidelines that clearly indicate to the private sector where it should be investing in order to develop supplies appropriate for DOD needs and national environmental policy standards.

Finally, over the long-term, *DOD should also consider fuel affordability: whether its supply systems will be able to operate for sustained periods of time without crippling negative direct costs and externalities.* In this sense, affordability applies to the actual cost of DOD's energy supplies and the risks that those supplies carry. This standard also indicates a need to consider the effects of potential price spikes on the defense budget—both within DOD if its fuel costs rise, and for the nation as a whole (if high prices negatively affect the economy in ways that lead to a constrained federal budget). Costs associated with the increasing difficulty in tapping the world's oil resources show that dependence on finite, nonrenewable resources is inherently risky. Indeed, the blanket assumption that petroleum would remain affordable indefinitely is what caused the dangerous dependence with which DOD now wrestles. It is critical, however, that affordability be considered with reference to the costs of fuels produced at scale. Any newly developed fuels that are not yet mass produced will cost more in their development stage and less once economies of scale are achieved. It will be incumbent on potential alternative fuel suppliers that their fuels will be affordable over the long term.

3. Plan for an Uncertain Future

DOD should forecast what its fuel vulnerabilities and needs are likely to be decades into the future as a means of guiding energy choices today. The future of DOD aviation and aviation fuel in particular will influence the pace and composition of DOD's energy strategy over the long term, considering that aviation fuel accounted for 56 percent of DOD's energy consumption as of 2008, and about 80 percent of the Air Force's energy needs.[33] The future of manned and unmanned flight will directly impact the balance of DOD's energy investments in jet fuel, energy storage devices and other energy technologies.

DOD should develop a series of planning scenarios to game out fuel needs against different potential future combat concepts. Warfare 20, 30 and 40 years from now will not look like today's wars. Likewise, the way the United States secures its interests will likely not mirror today's efforts. Preparing for this uncertainty requires thinking today about how DOD will operate years down the line—and this by necessity includes envisioning DOD's future energy needs.

The key to successful energy planning will be to ensure diversity within the scenarios: incorporating diverse needs and diverse sources of energy and the supply systems that they will require. Planning scenarios may blend together and overlap, but must involve planning for a very broad range of energy technologies and requirements. This will ensure that DOD is preparing for a wide range of energy contingencies. From these, it can derive estimates of what types of fuels, infrastructure and storage technologies it may need to invest in today.

---

[32] See GovTrack.us, Text of H.R. 6 [110th]: Energy Independence and Security Act of 2007.

[33] U.S. Energy Information Administration, *Annual Energy Review 2009*, "Table 1.13: U.S. Government Energy Consumption by Agency and Source, Fiscal Years 2003, 2008 and 2009." (19 August 2010).

4. Demand New Fuels for Old Equipment

The majority of the vehicles, aircraft and weapons systems that DOD purchases in the near term will be designed to be fueled by petroleum, as are most of DOD's current assets. Most of these systems will remain in commission for decades before replacements are seriously considered. Notably, the Office of the Secretary of Defense (OSD) is working to fulfill a mandate from Congress that defense suppliers work to increase fuel efficiency as a consistent part of acquisition processes. In the near term, DOD should also sustain its focus on dropin fuels—that is, liquid fuels designed as chemical equivalents to petroleum-based fuels, and that are therefore ready for immediate use in existing aircraft, vehicles and equipment once they are tested and certified. The Navy and Air Force have already begun moving down this path, and both have now flight-tested drop-in biofuels blended with petroleum-based jet fuel. The key will be to maintain and strengthen the demand signal these tests have begun to create in order to push the private sector to continue producing military-appropriate fuel supplies. It will also be important for DOD to continue to consider the long-term environmental ramifications of these drop-in fuels so as not to violate Congressional requirements that its alternative fuels have lower greenhouse gas emissions than petroleum equivalents.

Diversification of energy supplies stands to be an important benefit to DOD of this focus on drop-in fuels. Even if DOD positions itself to meet all of its energy needs using non-petroleum sources by 2040, there may still be circumstances in which certain fuels are simply not available when and where DOD needs them. If DOD can procure fuels from a portfolio of sources, such as fuels made from locally grown switchgrass, algae, camelina or other crops, that diversity can help to keep prices competitive (especially as a hedge against weather or economic conditions reducing crop output in any given region) and deny suppliers leverage over the United States. Diversification can also ensure that DOD will be able to procure the fuel it needs around the world. Enjoying the full operational and budgetary benefits of fuel diversification will also require DOD to work with foreign governments on international standards for military-grade fuels.

5. Continue to Increase Alternative Fuel Use at Domestic Installations

The best way to begin DOD's energy transition will be to begin with fast-tracked efforts at bases in the continental United States. The services are already increasing renewable power generation at their installations, and leaders at several bases have even set goals of becoming net-zero energy consumers (in other words, producing as much energy as they consume) and developing resilient microgrids. In several conversations with energy managers at U.S. bases during the course of our research, there was a tangible sense that increasing efficiency and use of renewable energy domestically contributed to the broader goal of DOD improving its long-term energy security.

To date, DOD has focused heavily on generating renewable electricity at domestic installations, but it should expand this focus to include reducing petroleum use in vehicle fleets. Moving to alternative fuels in ground vehicles will be easier than displacing aviation fuels, which require an array of additional specifications. At its installations, DOD also has more alternative fueling options that those designed for use in aviation (*e.g.,* DOD cannot fly its aircraft with electricity today, but it can adopt electric ground vehicles if they meet the guiding principles outlined above). This added flexibility allows individual bases to invest in energy sources that make sense given regional renewable energy production capabilities and infrastructure.

The Long Lives of DOD Assets

Given DOD's long acquisitions process, a majority of the vehicles, aircraft and weapons systems that DOD purchases in the near term will be designed to use petroleum-based fuels, as are most of DOD's current assets. Consider the following: for DOD's 2008 acquisitions programs, 27 of the 80 active programs had been in development for a decade or more. What is more, most of these systems will remain in commission for decades, and any DOD energy strategy will have to account for the fueling needs of these systems. Below are several programs, retired and active, that reflect DOD's long development and deployment timeline.

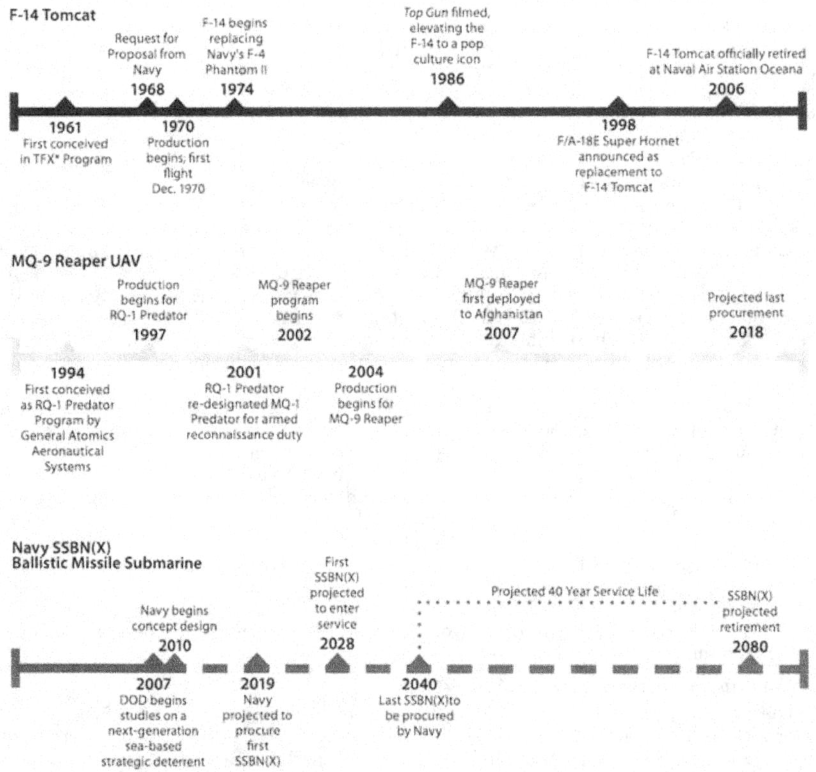

* Tactical Fighter Experimental program
Source: *Defense Acquisitions: Assessments of Selected Weapon Programs* (Government Accountability Office: Washington, D.C., March 2009):10.

6. Invest for Maximum Impact

DOD should maximize the impact of its investments by factoring distribution and infrastructure into its decisions on where to invest. Because energy production, purchasing, transport and transmission all involve systems of infrastructure and sunk costs, new fuels will not likely displace the old everywhere simultaneously. Prioritizing energy projects is today a bottom-up and organic process: Interested individuals navigate mazes of funding streams, laws, regulations, contract types and public utility relationships in order to gain approval and funding to move forward with renewable energy or efficiency projects. DOD should streamline this process and target it to maximize results.

For DOD to prioritize where to focus its energy transition efforts better, it should identify the locations where transition to non-petroleum fuels would have the greatest, most immediate impact. For example, DOD often uses jet fuel in vehicles and equipment due to the logistical benefits gained in using a single fuel type. Therefore, aviation fuel must be a central focus of this analysis. DOD should identify points at which drop-in biofuel blends or other energy systems will cover the greatest volume of fueling. As it considers this step, it will find the private sector aviation industry, which has considered prioritizing aviation biofuel supplies for the nation's busiest airports, instructive. For example, if biofuels are available at the seven busiest U.S. airports in passenger volume, they could power nearly 28 percent of the country's air traffic.[34] These airports could be used as hubs around which to

[34] According to the FAA, the top seven airports by passenger volume are:

Hartsfield—Jackson Atlanta International: 6.10 percent

Chicago O'Hare International: 4.45 percent

Los Angeles International: 3.94 percent

build energy infrastructure and production capacity in order to hasten the adoption of renewable fuels there. Cities around the United States and institutions such as the U.S. Postal Service have utilized their hub-and-spoke fueling systems to quickly integrate new fuels and vehicles into their fleets—cases which should be studied for best practices and important lessons in adopting new fuels.

Finding the locations with the greatest fuel demand, however, is only the first step since not all locations are currently conducive to the production, transport or use of non-petroleum fuels. DOD should therefore analyze the list of top fuel demand locations against key enablers that could hasten the availability of alternative fuels at a large scale. These enablers should include: permissive state and local laws and incentives; infrastructure to handle transport, storage and fueling; and supply availability (including states or regions with current biofuels development in progress or high production potential).

7. Save Energy, Keep the Change

Several disincentives hinder DOD's transition to more efficient energy use and the use of alternative fuels. The problems here run deep; over the course of research for this report we have heard from energy managers at U.S. military bases, installation policymakers in Washington and officers representing each of the services. Perhaps most importantly, individual bases, the military services and even the DOD writ large cannot always pocket and repurpose the money they save if their energy costs drop. This is a result of the type of funding used for renewable energy or efficiency investments or arrangements with local public utilities for renewable energy installations. Additionally, depending on how DOD pays for renewable energy investments on its bases, it does not always receive the commensurate clean energy credits for the energy generated on its land. These disincentives to save energy also extend to many contractors. Implementing a long-term energy strategy will therefore require DOD to address incentives and disincentives built into budgeting rules and norms, including for contractors. Energy Savings Performance Contracts, which allow contractors to recoup their energy investments in federal projects, are one example of how designing incentives for contractors to reduce energy use can dramatically lower consumption.[35]

Correlated to the current misalignment of incentives, DOD lacks appropriate metrics regarding its energy security activities. This stems in part from the lack of a long-term energy strategy or a specific, unified goal. OSD and the services do have long lists of metrics for meeting objectives that may or may not measure progress toward the endpoint DOD needs to reach. Past metrics have also tended to measure static energy use and do not account for military activities. New metrics to indicate DOD's success (or lack thereof) in progress toward its long-term energy goals should be both streamlined and meaningful.

8. Understand that Energy is Not Free

Changing how DOD meets its energy needs will involve a shift in its culture. It is important to note that this challenge is not distinct to DOD: Due to relatively (and often artificially) cheap energy and the normalization of consistent and abundant supplies, the country broadly undervalues the true cost of energy and therefore faces few incentives to change its behavior. Change will take time, and it will involve consistent leadership and public education. A culture that recognizes the cost of failing to change the energy status quo will help facilitate DOD's smooth transition to more sustainable long-term energy use. It will also have ripple effects for the country. Whether through disseminating new technologies such as GPS or leading by example to change cultural norms such as with racial integration, changes to DOD's culture often set the stage for significant national change.

Among those who consider DOD's energy challenges on a regular basis, a consensus has formed that cultural change is a necessary component of meeting long-term energy needs. One Marine Corps representative recently described DOD as a victim of its own success in that it manages logistics and engineering so well that energy is taken for granted: it is simply available when and where it is needed.[36]

Dallas/Fort Worth International: 3.83 percent

Denver International: 3.45 percent

John F Kennedy International: 3.26 percent

McCarran International: 2.79 percent

Total: 27.82 percent

[35] See Department of Energy, "Energy Savings Performance Contracts," (2010) for further information on ESPCs.

[36] Off-the-record CNAS event (July 2010).

The wars in Iraq and Afghanistan launched the process of reversing this trend, as supply lines have proven extremely vulnerable to attack.

Committed leaders are in place, meeting the first precondition for integrating energy into the normal ways in which DOD does business. Civilian and military leaders of the Navy, Marine Corps, Army and Air Force have all spoken to the importance of improving energy efficiency and assuring long-term fuel availability and created energy offices.

Next steps include raising awareness at every installation, and improving energy education at war colleges and through messaging by higher-ranking officers. The vast majority of representatives we spoke with at all civilian and military ranks during the course of this project understood the operational vulnerabilities involved with the high energy consumption required by the current wars. Subsequent areas of focus must include long-term energy supply and demand trends, the negative economic and environmental effects of fossil fuel dependence and trends in science and innovation.

9. Promote a Shared Vision of DOD's Energy Future

Even with all of DOD's efforts, it cannot meet its long-term energy goals without Congress, the rest of the Executive Branch and a critical mass of private companies sharing a similar vision. Businesses and academic researchers will have to do the heavy lifting in energy innovation, and DOD relies on Congress and the White House to provide funding. Yet while DOD has worked busily to define and confront its energy challenges over the past few years, its track record in relating its activities to the outside world is mixed at best. Many current successes are driven by individual initiative, making them ad hoc and easily terminated. Some aspects of external relations need major adjustment, while other areas of improvement will require relatively minor refinements.

Most critical is for DOD and Capitol Hill to improve communication on energy issues. Legislators and their staffs often are left to interpret for themselves what energy policies it would be helpful to require for DOD. Many at DOD also express frustration that energy requirements mandated by Congress are not always backed by funding to invest in steps like fuel switching, new infrastructure and efficiency upgrades. DOD should develop a robust plan for Capitol Hill relations and external relations to communicate its long-term energy strategy. It should ensure that its strategic thinking is framed clearly and points toward real policy actions that Congress (or other government agencies) can adopt. There is also a strong need for Congressional staffers to expand their knowledge on DOD energy issues, and to ensure due diligence in examining how DOD may react to their ideas before they are enacted in law.

A simple way for DOD to improve its relations with other government agencies is to provide an online organization chart of major DOD offices focused on energy and a description of the general roles and responsibilities of those offices. This may seem simplistic, but to those not familiar with the DOD bureaucracy (especially policymakers on the Hill and clean energy entrepreneurs) it can be extremely challenging to find the proper points of contact to discuss energy policies in DOD. There is little hope of improving interagency coordination or Congressional relations if outsiders cannot even figure out whom to engage with questions or ideas.

10. Engage Allies in the Energy Transition

Through foreign military sales, joint exercises and international basing, DOD can promote adoption of shared technical standards and directly influence the energy systems used by its allies. This will improve its own ability to operate by ensuring that the United States has access to needed energy supplies globally and improving interoperability. It will also encourage allies to make compatible choices with respect to energy, instead of working at cross purposes.

DOD's long-term energy strategy must therefore include an international plan of action. At a minimum, this should include information sharing on alternative energy research and development. It should also include cooperation with international partners on fuel testing and evaluation, and setting fuel standards that guarantee interoperability. This should be a familiar concept for DOD, which already sets joint standards with allies by, for example, standardizing the use of 9mm NATO cartridges by all member countries. Where the interests and regulations of both countries permit, such efforts can include working with U.S. allies on energy technology sharing. This will also require better coordination with the State Department, the National Labs and U.S. energy industries. Additionally, it will have the positive effect of signaling to international suppliers (both countries and private companies) that DOD will favor procurement of non-petroleum fuels when possible.

Energy is an increasingly important issue for U.S. diplomacy with traditional allies such as Japan, the Republic of Korea (ROK) and NATO countries. Where these overlap with important military considerations, DOD's active engagement will be critical for ensuring that its needs are considered. But while this step may seem straightforward and relatively easy to implement, in fact each country has its own interests, domestic politics, economic pressures and tradeoffs to consider. Often, logical areas of cooperation on energy are in fact areas of competition. Cooperation regarding installation energy use can be particularly difficult as it is often met with requirements that favor American products.

It is important to remember that DOD already works internationally to secure its energy supplies for the current petroleum-heavy system—and that the process is often neither smooth nor easy. Contracting and using supply systems for petroleum through countries such as Azerbaijan are already costly and require often-difficult relationship management.[37] DOD should actively consider how it can better coordinate with U.S. allies to develop nonpetroleum energy systems to meet its requirements for reliable, affordable and sustainable fuels.

11. Streamline Energy Management

Managing a smooth transition from petroleum to meet DOD's long-term energy needs will require bureaucratic and personnel changes. DOD's current structure reflects past thinking about energy rather than current priorities, and the military services and OSD regularly change the structure of their offices and personnel requirements to address questions of energy. Energy planning and policy are also subject to changing mandates by Congress and the White House. It is important to underscore that many aspects of DOD's current energy personnel structure mark major improvements and indicate solid leadership on energy. As mentioned earlier, each branch of the armed services and OSD have new offices devoted specifically to energy, including experts on operational fuel use. Nonetheless, incorporating energy better into how DOD does business, as the 2010 QDR mandates, is far from institutionalized.

Figure 1: Recommended Personnel Structure, Office of the Secretary of Defense

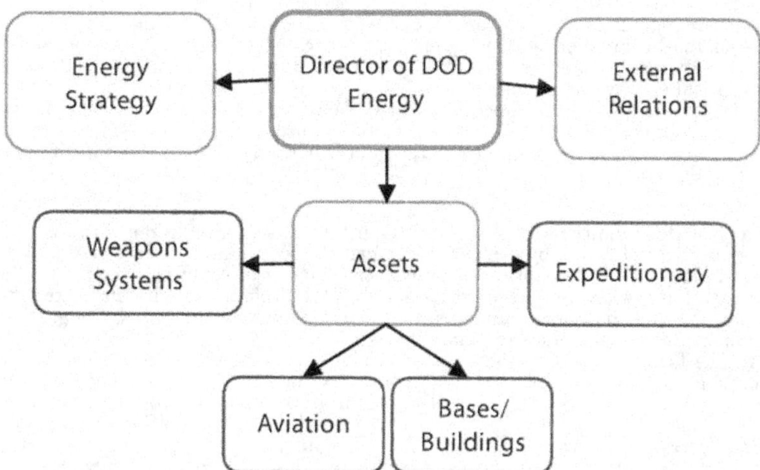

Within OSD and the services, responsibility is generally split between those managing energy for military installations and those managing operational energy. This is in part a legacy divide: Positions governing operational energy in OSD and the services have only been stood up as dedicated offices over the past few years, while offices governing energy use at military bases have long been part of the DOD organizational structure. However, the separation of energy along these lines is a false distinction; the training and equipping carried out at domestic military installations

[37] Craig Whitlock, "Gates brings reassurances to Azerbaijan leader," *The Washington Post* (7 June 2010).

is geared toward operational utility. The only truly static component of installations is the buildings themselves, whereas the people using energy and how they use it fluctuate regularly and depend on operational requirements. Indeed, the definitions of operational and non-operational energy are not well delineated in related laws or Congressional requirements.

Once a long-term DOD energy strategy is in place, the DOD should assess its related organizational and personnel structures. This assessment should involve an evaluation of personnel needs, and in particular what positions are filled by political appointees versus civil service officers versus contractors, while being cognizant of the work that the military services themselves conduct.

Since the separation of installation and operational energy reflects DOD's energy past more than its energy future, it should seek in the years ahead to merge energy management at the OSD level to a coherent body under the leadership of one individual. The Army, Air Force, Navy and Marine Corps could continue to manage their own unique energy bureaucracies as their leaders deem best. This combined office should include experts focused on the following important areas:

Strategy

A major component of DOD's energy strategy will include setting priorities, planning against various scenarios and contingencies, and tracking progress against objectives.

Assets

One person should oversee energy issues as related to specific DOD assets, with a team of individuals focused specifically on the very different categories of assets. This component would consider not just the stock of DOD equipment, vehicles, ships and aircraft, but also long-term trends in how DOD employs them.

Aviation: As it comprises more than half of DOD's petroleum use and requires unique technical knowledge, aviation fuel is a category onto itself.

Weapons Systems: Assets such as missile defenses and directed energy weapons also have unique energy signatures and, given their limited numbers and specific uses, are operated differently from other categories of assets. Parsing which weapons systems have unique enough energy requirements to necessitate consideration independent of the expeditionary and aviation categories will be difficult, and they will likely change over time.

Expeditionary Energy: This component would include all mobile assets not represented in the aviation and weapons systems categories. It will be the heart of DOD energy activities during wartime, when fuel to deployed troops represents the most critical energy management.

Buildings/Bases: This component of DOD's energy infrastructure should focus only on installations themselves. It will require coordination with public utility commissions and legal and regulatory bodies, and knowledge of often-complicated state and local dynamics.

External Relations

Many of the conditions that will determine DOD's ability to meet its long-term energy needs will be set by Congress, the private sector and the international community. Meeting DOD's energy needs over the long term requires effective relations with all of these groups. This component will therefore include three important areas of external relations management: private sector partnerships; Congressional relations; and international relations.

Officials focusing on all of these areas will be responsible for interagency coordination and coordination within OSD as it relates to their work. As much of the activity on meeting energy goals does and will continue to reside among the services, coordination among them and by OSD will be imperative. These positions will also represent a straightforward network of points of contact for other government and non-governmental representatives needing to coordinate with DOD on energy issues.

Funding for DOD's investments in reliable long-term energy supplies will come in many forms, and it will be critical for DOD's energy personnel to develop a deep understanding of how to properly resource its energy strategy. New resources should go toward sunk costs—efficiency upgrades, fuel testing and evaluation and energy infrastructure.

However, meeting DOD's goal of making a smooth transition away from petroleum will require the private sector to provide cost-competitive, at-scale renewable fuels that the Defense Logistics Agency can purchase when and where it needs them. This will require DOD to commit to a general direction for its energy future in order to send an effective market signal, and it will require incentives and regulations beyond DOD's control.

Contracting mechanisms and direct funding appropriated by Congress will constitute important means for making the necessary sunk investments for renewable energy adoption. The 2009 American Recovery and Reinvestment Act proved to be a successful and popular stream of funding for several projects at domestic installations, and lessons learned can be collected to indicate where funding may be most effective for future projects. The services also devote significant resources toward meeting this challenge. The Navy and Air Force have been testing and certifying alternative aviation fuels within their own budgets. They will need to remain consistent in these investments for some time, but the rewards in potential savings to their budgets should over the long-term pay off if DOD can properly align its incentive structures.

Given the urgent need to address operational energy considerations in the current wars, this grand bureaucratic adjustment might best be timed for after significant redeployments from Iraq and Afghanistan are complete. Managing DOD's long-term energy transition may not need a vast personnel structure in its next iteration, though each component of the office can grow or shrink to match the changing nature of DOD's activities. The Army, Navy, Marine Corps and Air Force are also likely to continue to provide personnel who will address energy challenges, and much implementation will be conducted by base managers.

12. Plan for the Worst

DOD should plan for contingencies in which its predictions and plans for moving beyond petroleum turn out to be wrong. In other words, its "off-ramps from petroleum" may turn out to be rough roads, or DOD could make the wrong turns or miss the ramps altogether. For instance, DOD should imagine scenarios involving absolute shortages of energy, major price spikes, alternative fuels that simply cannot scale up fast enough and major technological or environmental game-changers that fundamentally alter how DOD meets its energy needs.

If worst-case scenarios transpire, they could cost DOD its ability to operate effectively. DOD, including the war colleges, combatant commands and OSD, has already conducted war games and scenario exercises that include fuel shortages, extended blackouts and other contingencies. DOD must continue to think through these kinds of scenarios, compile lessons learned from them and apply them to its energy calculations.

IV. Conclusion

The steps outlined in this report will help DOD transition to non-petroleum sources of energy, to the benefit of national security and operational effectiveness. Yet DOD's smooth transition to a future energy paradigm that does not rely on petroleum depends heavily on policies that lie beyond its own control. Many relevant policy choices and commitments are up to elected officials, state and local governments, the private sector and the international community (see Appendix II: *How the Rest of the Government Can Contribute to DOD's Energy Strategy*). Congress and the White House will continue to refine energy requirements for all federal agencies, and exert their leadership to improve the American public's understanding that these actions are taken to promote U.S. national security. DOD's long-term energy strategy should include coordination with all these groups, since their decisions will affect DOD's ability to operate.

Meeting DOD's energy demands with new fuel sources in the next 30 years will require patient and persistent leadership by DOD officials. But the benefits will prove to be far-reaching. These changes will help DOD to hedge against unbearable costs, maintain its flexibility and guarantee its ability to protect and defend the United States against all enemies—regardless of the availability of petroleum-based fuels.

Appendix I : Why Examine Reserve-to -Production Ratios?

By Alexandra Stark

The U.S. Energy Information Administration defines the R/P ratio as "the number of years that oil and gas reserves would last at the current production rate." The resulting figure indicates the length of time in years that known, recoverable reserves are expected to last if production continues at the same pace. This timeline, while constantly in flux, gives a more useful indicator than just supply, demand or reserve figures for the purpose of long-term policy planning.

Global proved reserves (the quantities of oil that exist with reasonable certainty and can be recovered under current geological, economic and technological conditions) are often cited in considering the future of world energy trends. However, this is not always a helpful indicator. For example, Saudi Arabia has almost 20 percent of remaining global proved reserves, but at the current rate would produce its reserves in less time than Venezuela, which has about 13 percent of reserves but produces those reserves much less efficiently. Ominously, many major suppliers to the United States could produce their current proved reserves in fairly short time horizon if they continue at the present rate: For example, the R/P ratio for Canada (the top supplier to the United States in 2009, providing more than 20 percent of total oil imports) stands at about 28 years today. For the United States itself, it is 11 years. The only countries with current R/P ratios longer than 75 years are Venezuela, Iran, Iraq, Kuwait and the United Arab Emirates.

Figure 2: World Petroleum Reserve-to-Production Ratios

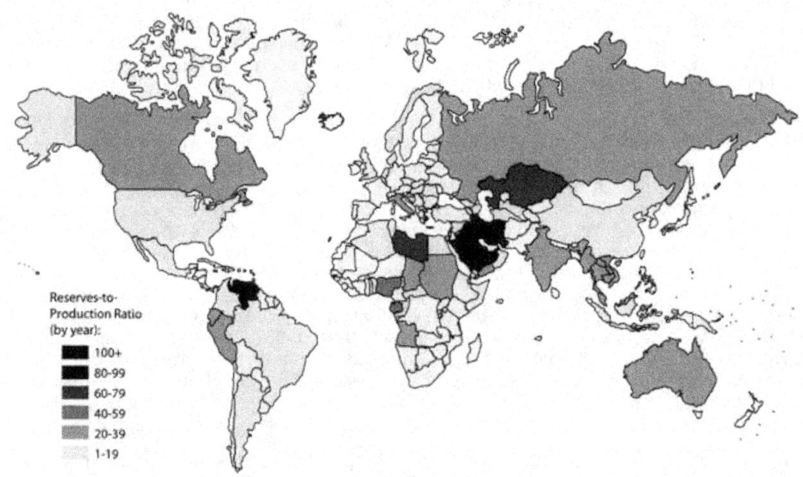

Source: BP Statistical Review of World Energy 2010
Sources: BP Statistical Review of World Energy 2010 and U.S. Energy Administration, "U.S. Imports by Country of Origin." Data are from 2009.

It is also important to note that both elements of this ratio change regularly, so even reserveto- production ratios are not perfect predictions of the future. As prices change and technology advances, as demand rises and falls, and as new reserves become accessible, R/P ratios are likewise affected. However, just as it is possible for countries to have a longer time horizon than projected, it is also possible for countries to exhaust their reserves sooner than expected, and for rising prices to make non-petroleum fuels more cost-competitive than investing in new petroleum production. While no economic and geologic estimates perfectly predict the future, R/P ratios can serve as important indicators for DOD officials and policymakers to plan against.

APPENDIX II: HOW THE REST OF THE GOVERNMENT CAN CONTRIBUTE TO DOD'S ENERGY STRATEGY

Many of the policies and measures that will help DOD achieve its long-term energy goal of making a smooth transition away from petroleum by 2040 lie beyond DOD's jurisdiction. The following actions by Congress, the White House and the private sector will contribute to DOD's continued ability to meet its energy demands within the constraints outlined in this report.

Provide a clear long-term legal and regulatory environment. Market-based regulatory adjustments and innovation coming holistically through the private sector will be more helpful than DOD pushing for different systems piece by piece. Unfortunately, today many businesses are biding time and waiting for a more certain business environment rather than producing the fuels they have developed and making the investments they have planned. Hundreds of businesses have encouraged the Federal Government to pass clean energy and climate change legislation to provide a significant long-term market signal. Doing so should be considered one of the primary ways that the Nation's leaders can help ensure that DOD can meet its long-term energy needs.

Mind the grid. DOD's ability to address its electricity reliability concerns is in part beyond its own jurisdiction. Almost all of DOD's domestic installations are connected to the public power grid and must therefore rely on local or regional utilities to grant it permission for renewable energy production and to improve grid reliance. The utilities are working to bolster grid security, but concerns remain sufficient that many at DOD and in Congress are considering plans for "islanding" bases, or detaching them from the public grid system altogether. Public utilities should continue to work closely with nearby installations to ensure that public and defense community needs are taken into account. A consistent legal and regulatory environment would also promote decisions by utilities to make investments in new infrastructure and rules to allow greater renewable energy production.

Extend requirements from Congress. The 2007 Energy Independence and Security Act (EISA) requires federal buildings, including domestic DOD installations, to reduce energy consumption up to 30 percent through 2015. This raised the bar from previous requirements set in 2005. Congress should direct additional requirements for efficiency and use of renewable energy in domestic installations beginning when previous requirements are set to end (often 2015). It should also continue to mandate that the fuels that federal agencies invest in have lower greenhouse gas emissions than the fuels they are meant to displace. However, two changes may be in order. The 2007 legislation requires that DOD reduce energy per square foot, yet this calculation does not account for the dramatic differences in the ways in which DOD uses different facilities. Congress should also be sensitive to the tight budget environment that DOD officials feel, and consider prizes for innovation and other mechanisms to provide funding to meet these requirements. The next round of legislative change to require DOD's continued progress on energy should be designed through extensive discussions and good coordination between DOD and the Hill.

Address information challenges. Credible government estimates are available for fossil fuel resources, including specific estimates of energy reserves, production, consumption and historical prices. These include reserve-to-production projections and future outlooks that are generally reliable, if often conservative. Finding comparable information for non-fossil fuels is difficult to impossible, and often involves wading through dense reports. There is no single-source place where those reports lie, and analysts are left to compare and judge the efficacy of sources on their own. The private sector often provides more accessible information—but not information that can necessarily be relied upon as neutral and accurate. While we do not recommend that the Federal Government engage in guesswork or estimates that are less than diligent, DOD must recognize this information gap.

Make reliable models available. DOD's incorporating greenhouse gas emissions, economic costs and other lifecycle effects of its energy options presents its own challenges. The computer models used to make these calculations reflect the sum of their parts: the data and mechanisms used by the modelers must be accurate (and reflect honest scientific facts, not political agendas or skewed information) to produce viable calculations. Information on the carbon, water and land use footprints of emerging fuel sources can also be more difficult to calculate than those of long-used sources, as they suffer from relevant information often being proprietary and in the hands of private companies. New fuels may also be adaptable to meet specific environmental footprint requirements once they are developed and produced at scale, which is a positive factor but again difficult to quantify. Meeting environmental constraints can be an inexact science, and calculations can change over time. DOD should therefore rely on energy and related climate models run by or compared to honest brokers, such as academics or the National Labs, in its decision-making.

○

www.ingramcontent.com/pod-product-compliance
Lightning Source LLC
Chambersburg PA
CBHW081548170526
45166CB00009B/2624

* 9 7 8 1 4 7 8 3 1 2 3 6 9 *